S0-ARM-681

Issues, Evidence and You

Issues, Evidence and You

SCIENCE
EDUCATION FOR
PUBLIC
UNDERSTANDING
PROGRAM

LAWRENCE HALL OF SCIENCE LHS
UNIVERSITY OF CALIFORNIA AT BERKELEY

RONKONKOMA, NEW YORK

This book is part of SEPUP's middle school science course sequence:

Science and Life Issues	Issues, Evidence and You
My Body and Me	Water
Micro-Life	Materials
Our Genes, Our Selves	Energy
Ecology and Evolution	Environment
Using Tools and Ideas	

Additional SEPUP instructional materials include:

CHEM-2 (Chemicals, Health, Environment and Me): Grades 4–6

SEPUP Modules: Grades 7–12

Science and Sustainability: Course for Grades 10–12

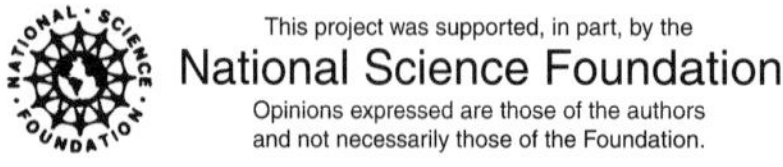
This project was supported, in part, by the
National Science Foundation
Opinions expressed are those of the authors
and not necessarily those of the Foundation.

3 4 5 6 7 8 9 08 07 06

ISBN: 1-887725-70-9

Editor: Mary Jean Haley
Cover, text design and production: Seventeenth Street Studios
SEPUP publication review: Miriam Shein

SEPUP
Lawrence Hall of Science
University of California at Berkeley
Berkeley CA 94720-5200

e-mail: sepup@uclink4.berkeley.edu
Website: www.sepuplhs.org

Published by:

17 Colt Court
Ronkonkoma NY 11779
Website: www.lab-aids.com

A Letter to IEY Students

As you examine the activities in this book, you may wonder, "Why does this book look so different from other science books I've seen?" The reason is simple: it is a different kind of science program, and only some of what you will learn can be seen by leafing through this book!

Issues, Evidence and You (IEY) uses several kinds of activities to teach science. For example, you will design and conduct an experiment to determine the amount of acid in simulated industrial waste water. You will investigate the chemical and physical properties of plastics and other common materials. And you will apply what you have learned about energy and materials to design an energy-efficient car.

You will find that important scientific ideas come up again and again in different activities. You will be expected to do more than just memorize these concepts; you will be asked to explain and apply them. In particular, you will improve your decision-making skills, using evidence and weighing outcomes to decide what you think should be done about scientific issues facing society.

How do we know that this is a good way for you to learn? In general, research on science education supports it. In particular, the activities in this book were tested by hundreds of students and their teachers, and they were modified on the basis of their feedback. In a sense, this entire book is the result of an investigation; we had people test our ideas, we interpreted the results, and we revised our ideas! We believe the result will show you that learning more about science is important, enjoyable, and relevant to your life.

SEPUP Staff

Acknowledgments

SEPUP STAFF

Dr. Herbert D. Thier, *Program Director*
Dr. Barbara Nagle, *Co-Director*
Janet Bellantoni, *Instructional Materials Developer*
Manisha Hariani, *Instructional Materials Developer*
Paul Hynds, *Instructional Materials Developer*
Laura Kretschmar, *Instructional Materials Developer*
Daniel Seaver, *Instructional Materials Developer*
Dr. Marcelle Siegel, *Instructional Materials Developer*
Marlene Thier, *Teacher Education and CHEM Coordinator*
Dr. Peter Kelly, *Research Associate (England)*
Dr. Magda Medir, *Research Associate (Spain)*
Mike Reeske, *Development Associate*
Miriam Shein, *Publications Coordinator*
Roberta Smith, *Administrative Coordinator*
Donna Anderson, *Administrative Assistant*

CONTRIBUTORS TO THIS BOOK

This edition of *Issues, Evidence and You* represents an adaptation of previous material. Instructional review and revision were performed by Mark Koker, Herbert D. Thier, and Barbara Nagle.

Major contributors to the original development and writing of *Issues, Evidence and You* were:

Robert Horvat
Barbara Nagle
Mike Reeske
Stephen Rutherford
Herbert D. Thier

Significant contributions were also made by Pamela T. Boykin, Mark Koker, and Marcelle Siegel (research).

Teacher participants in the original summer writing and editorial conferences were:

Eddie Bennett, *New York, NY*
Kathaleen Burke, *Buffalo, NY*
Carolyn Delia, *Marquette, MI*
Cynthia Detwiler, *Louisville, KY*
Richard Duquin, *Buffalo, NY*
Anne Little, *Winston-Salem, NC*
Donna Markey, *Oceanside, CA*
Susie Nally, *Lexington, KY*
Lori Sheppard-Gillam, *Anchorage, Alaska*
Linda Sherrill, *Sand Springs, OK*
Dorothy Trusclair, *Baton Rouge, LA*

The assessment system was developed in collaboration with the Berkeley Evaluation and Assessment Research (BEAR) Group at the Graduate School of Education, University of California, Berkeley. Major contributors included:

Robin Henke
Lily Roberts
Kathryn Sloane-Weisbaum
Chris White
Mark Wilson

Scientific review of the original *IEY* material was provided by:

Gary Arant, *General Manager, Valley Center Municipal Water District*
Joe Davis, *retired science teacher*
Rollie Myers, *Chemistry Department, University of California at Berkeley*
Steve Ruis, *Chemistry Department, American River College*

PROGRAM DEVELOPMENT CENTERS—DIRECTORS AND TEACHERS

The classroom is SEPUP's laboratory for development. We are extremely appreciative of the following center leaders and teachers who taught the program. These teachers, and their students, contributed significantly to improving the scope, quality, and teachability of the course.

Alaska: Donna York (Director), Kim Bunselmeyer, Linda Churchill, James Cunningham, Patty Dietderich, Lori Gillam, Gina Ireland-Kelly, Mary Klopfer, Jim Petrash, Amy Spargo

California-San Bernardino County: Dr. Herbert Brunkhorst (Director), William Cross, Alan Jolliff, Kimberly Michael, Chuck Schindler

California-San Diego County: Mike Reeske and Marilyn Stevens (Co-Directors), Pete Brehm, Donna Markey, Susan Mills, Barney Preston, Samantha Swann

California-San Francisco Area: Stephen Rutherford (Director), Michael Delnista, Cindy Donley, Judith Donovan, Roger Hansen, Judi Hazen, Catherine Heck, Mary Beth Hodge, Mary Hoglund, Mary Pat Horn, Paul Hynds, Margaret Kennedy, Carol Mortensen, Bob Rosenfeld, Jan Vespi

Colorado: John E. Sepich (Director), Mary Ann Hart, Lisa Joss, Geree Pepping-Dremel, Tracy Schuster, Dan Stebbins, Terry Strahm

Connecticut: Dave Lopath (Director), Harald Bender, Laura Boehm, Antonella Bona-Gallo, Joseph Bosco, Timothy Dillon, Victoria Duers, Valerie Hoye, Bob Segal, Stephen Weinberg

Kentucky-Lexington Area: Dr. Stephen Henderson and Susie Nally (Co-Directors), Stephen Dilly, Ralph McKee II, Barry Welty, Laura Wright

Kentucky-Louisville Area: Ken Rosenbaum (Director), Ella Barrickman, Pamela T. Boykin, Bernis Crawford, Cynthia Detwiler, Denise Finley, Ellen Skomsky

Louisiana: Dr. Sheila Pirkle (Director), Kathy McWaters, Lori Ann Otts, Robert Pfrimmer, Eileen Shieber, Mary Ann Smith, Allen (Bob) Toups, Dorothy Trusclair

Michigan: Phillip Larsen, Dawn Pickard, and Peter Vunovich (Co-Directors), Ann Aho, Carolyn Delia, Connie Duncan, Kathy Grosso, Stanley Guzy, Kevin Kruger, Tommy Ragonese

New York City: Arthur Camins (Director), Eddie Bennett, Steve Chambers, Sheila Cooper, Sally Dyson

North Carolina: Dr. Stan Hill and Dick Shaw (Co-Directors), Kevin Barnard, Ellen Dorsett, Cameron Holbrook, Anne M. Little

Oklahoma: Shelley Fisher (Director), Jill Anderson, Nancy Bauman, Larry Joe Bradford, Mike Bynum, James Granger, Brian Lomenick, Belva Nichols, Linda Sherrill, Keith Symcox, David Watson

Pennsylvania: Dr. John Agar (Director), Charles Brendley, Gregory France, John Frederick, Alana Gazetski, Gill Godwin

Washington, D.C.: Frances Brock and Alma Miller (Co-Directors),Vasanti Alsi, Yvonne Brannum, Walter Bryant, Shirley DeLaney, Sandra Jenkins, Joe Price, John Spearman

Western New York: Dr. Robert Horvat and Dr. Joyce Swartney (Co-Directors), Rich Bleyle, Kathaleen Burke, Al Crato, Richard Duquin, Lillian Gondree, Ray Greene, Richard Leggio, David McClatchey, James Morgan, Susan Wade

Contents

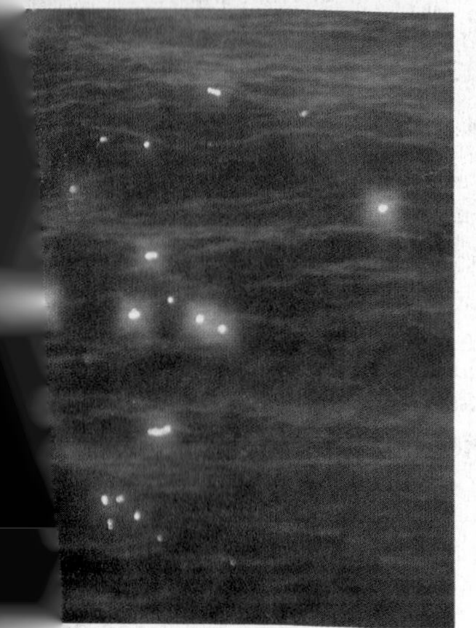
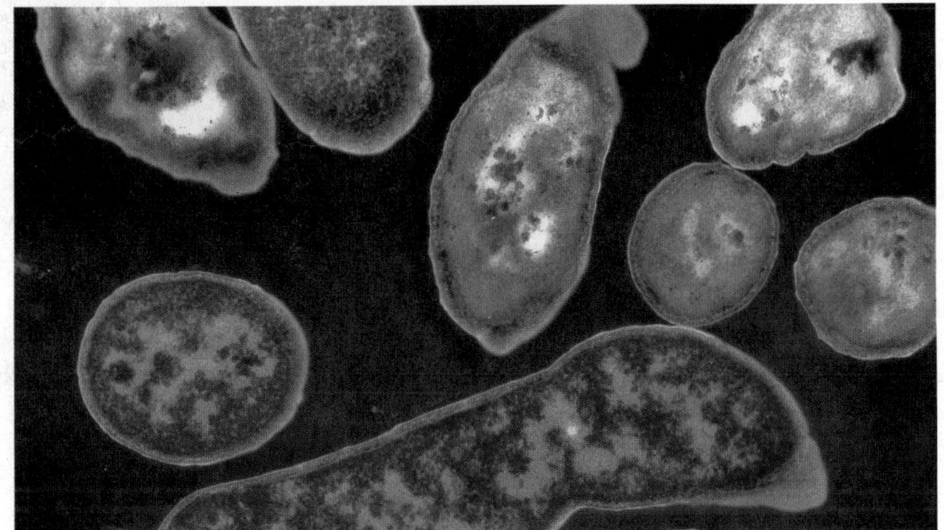

Water

1 Water Usage and Safety

Water is essential for life as we know it. This is as true for the plants and animals that live on land as it is for those that live in water. One of the most important reasons for this is that most substances on earth dissolve in water, at least to some extent. Our blood, which is mostly water, carries the oxygen we breathe and the dissolved nutrients from our food to every cell in our bodies. It also carries out the dissolved waste products.

Because of the important role water plays in our lives, the first six activities in this book explore the primary health risks associated with poor water quality. How do you know the water you drink is pure and safe? What does your community do to make sure its water is safe to drink? Whose responsibility is it? Activities 7 through 13 examine solution chemistry and chemical testing—what happens to substances once they have dissolved in water, and how we can use chemical tests to tell whether contaminants are present in water.

As you investigate these questions and discuss the evidence you gather in the activities in Part 1, you will be learning and practicing skills you need for the activities in Part 2.

1 Drinking Water Quality

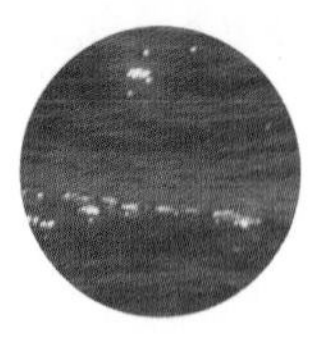

Can you taste the difference?

You have probably tasted bottled spring water. Every year in the United States, people spend billions of dollars on bottled water. Do you think people can tell the difference between tap water, distilled water (water without dissolved solids), and bottled spring water?

CHALLENGE

Your challenge is to find out if you or your classmates can identify the more expensive bottled spring water in a taste test.

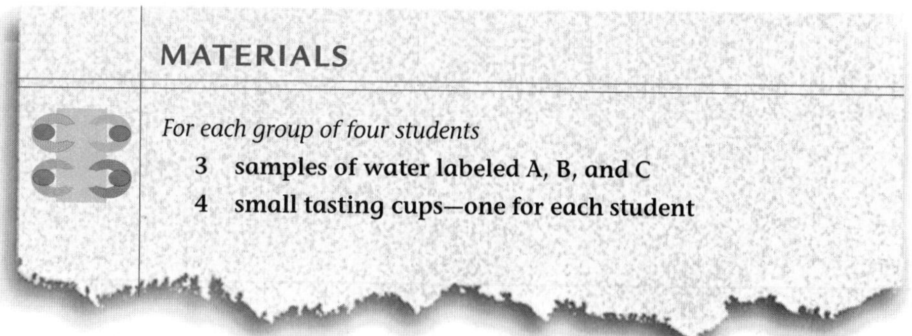

MATERIALS

For each group of four students

3 samples of water labeled A, B, and C

4 small tasting cups—one for each student

SAFETY NOTE: Use your own tasting cup. Remember, taste chemicals in the lab only when your teacher tells you to do so.

PROCEDURE

1. Set up your investigation report according to your teacher's instructions.
2. Fill each person's tasting cup half full of water sample A.
3. Observe the appearance of the water sample. Smell the sample. Finally, taste the sample. Drink all of the water in the cup.
4. Record your observations of appearance, smell, and taste in your science notebook. Make a larger version of the data table shown in the sample below. Do not share your results with other members of the group.
5. Repeat Steps 2 through 4 for water samples B and C.

Water Tasting (reproduce in your science notebook)

	Observations			
Water Sample	Appearance	Smell	Taste	Proposed Sample Identity
A				
B				
C				

6. Identify the sample that you think is the bottled spring water. Record your result in your data table.

7. When everyone in the group is finished, have each person share his or her results with the group. Everyone should give a reason for identifying one of the samples as bottled water.

8. The group should now discuss their reasons and try to reach an agreement about which of the samples is spring water. Record the result in a data table like the one shown below.

GROUP CHOICE for identity of bottled spring water

REASON ______________________________

____

____

CLASS CHOICE for identity of bottled spring ______________

REASON ______________________________

ACTUAL SAMPLE IDENTITY

SAMPLE A ______________________________

SAMPLE B ______________________________

SAMPLE C ______________________________

ANALYSIS

Record in your science notebook the answers to these questions. Follow your teacher's instructions on the format for your answers.

1. Did the spring water sample taste best to you?

2. Would you spend the extra money on spring water after the taste test? Why or why not?

3. What other information about bottled, tap, and distilled water would you like to know before you decide which water to drink?

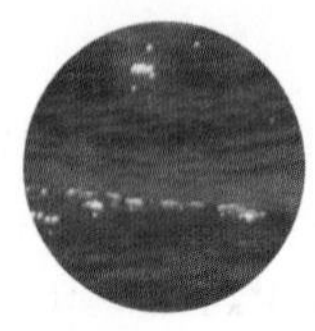

Purity in a Bottle

Do you think bottled spring water is worth the extra money? Is there more information you need to know before you decide? Sometimes bottled spring water may not be what you think it is.

CHALLENGE

Read the following and decide if it is a fair treatment of the issue: "Should I drink bottled water?"

PURITY IN A BOTTLE

Millions of Americans drink bottled water instead of—or in addition to—their local tap water. Surveys suggest than half of Americans drink bottled water and about a third consume it regularly. Sales have tripled in the last ten years, to over $5 billion a year. By the year 2005, the U.S. market will be worth about $7.3 billion.

How do Americans compare with the rest of the world? The world bottled water market represents an annual volume of 89 billion liters, and is estimated to be worth over US $22 billion. No one drinks more than the Italians (107 liters per year per person) and Western Europeans, as a whole, drink nearly half of the world's bottled water.

Why do people drink bottled water? Most do it for reasons related to health, taste, or convenience. It's hard to argue taste and convenience, since these are personal preferences. But a study released in May 2001 says that despite selling for up to 1,000 times the price of tap water, bottled water may be no safer. And, when you add in the additional packaging and energy needed to get the water to market, the impact of bottled water on the environment may be higher. More than 1.5 million tons of plastic are used to bottle water annually, and one-fourth of the bottled water is consumed outside its country of origin.

Yet the only difference between some bottled waters and tap water is that one is distributed in bottles, the other through pipes. Until 1993, such terms as "mineral," "spring," "artesian," "well," "distilled," and "purified," were used on labels without a standard meaning. These terms could be used if only 10% of the water came from that source. People buying bottled water with pictures of high mountain lakes on the label were often dismayed to find that their bottles contained mostly tap water! In the early 1990s, the U.S. House Energy and Commerce Committee reported that 25% of bottled waters came from the same source as local tap water. Current industry and government estimates suggest that 25–40% of bottled water still comes from tap water, sometimes with additional treatment, sometimes not.

The Food and Drug Administration (FDA) rules do not apply to 60–70% of the bottled water sold in American because its rules do not apply to water bottled and sold in the same state. And the United States Environmental Protection Agency (EPA) has different standards for tap and bottled water. Tap water must be tested for harmful bacteria more than 100 times per month, while bottled water plants must test only once a week. The standards governing nearly 200 inorganic and organic chemicals are less strict for bottled water. And states in general have far fewer staff assigned to routinely test bottled water than tap water—in some cases only one or two persons for the entire state, while hundreds of federal and state workers oversee the safety of tap water.

Supporters of bottled water say that there are few recorded instances of someone getting sick from bottled water. Further, the EPA reported in

1996 that about 1 in 10 community tap water systems—serving about 1/7 of the U.S. population—violated tap water treatment standards. Tap water supporters are quick to point out that these are "attainment" standards—future goals for increased water safety, and that tap water is safe.

QUESTIONS

1. What is the point of view expressed in the reading?

2. What evidence is included in the reading? (Evidence generally refers to facts or observations that can be tested or checked.)

3. Do you agree or disagree with the reading? Cite evidence to support your position.

4. What additional evidence would you like before deciding which water to drink?

2 Exploring Sensory Thresholds

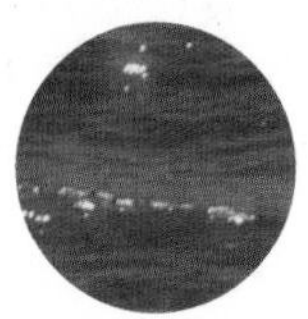

In the last activity you tried to identify bottled spring water by taste. In this activity, you will investigate the ability of two more senses—sight and smell—to detect substances dissolved in water.

CHALLENGE

Determine which one of your senses—taste, smell, or sight—can detect the lowest concentration of a drink mix solution. (Here, a concentration means the amount or proportion of a substance dissolved in the water.) Use a bar graph to compare the range of these senses for all the members of your class.

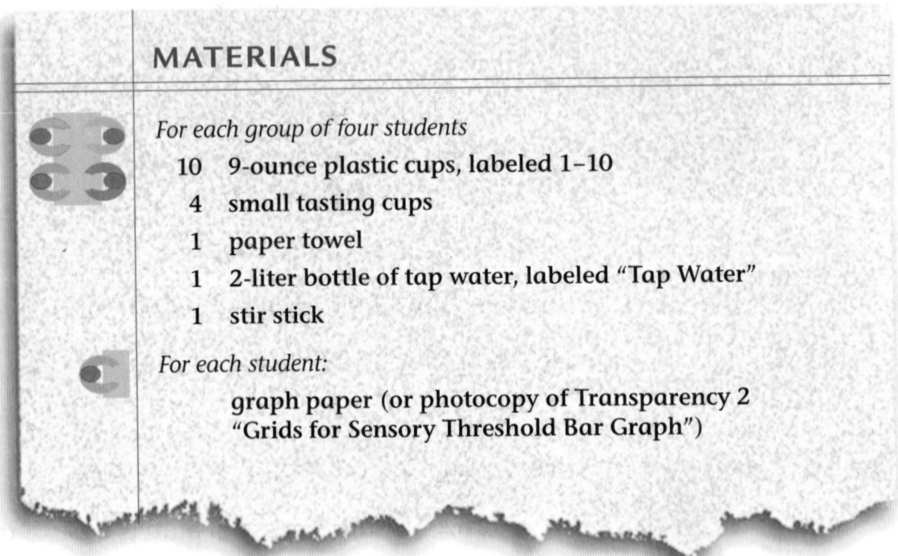

MATERIALS

For each group of four students

10 9-ounce plastic cups, labeled 1–10
4 small tasting cups
1 paper towel
1 2-liter bottle of tap water, labeled "Tap Water"
1 stir stick

For each student:

graph paper (or photocopy of Transparency 2 "Grids for Sensory Threshold Bar Graph")

SAFETY NOTE: Use clean plastic cups for this investigation. Do not share the tasting cups, and do not add anything to the plastic cups, including your fingers. Remember, taste chemicals in science class only when your teacher tells you to do so.

PROCEDURE

Part One

1. Set up your investigation report.
2. Based on your own personal experiences, predict which of your senses—sight, smell, or taste—can detect the lowest concentration of drink mix dissolved in water. Record your prediction. In your science notebook, give two reasons for your choice.
3. Your teacher will fill Cup 1 half full of concentrated drink mix solution.
4. In your group fill the cup to the 200-mL mark with tap water from the bottle. Mix with the stir stick.

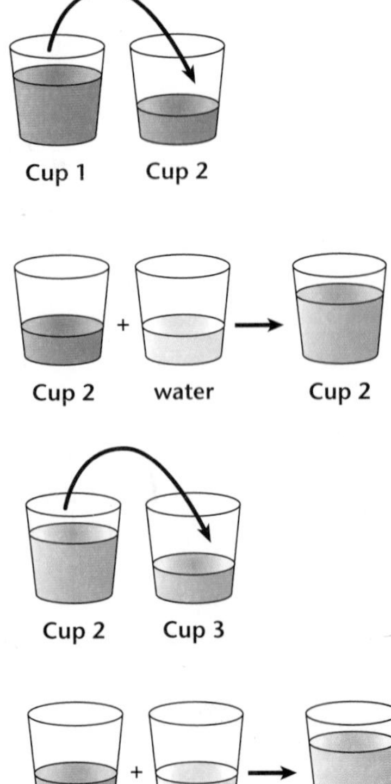

5. Pour half of the liquid from Cup 1 into Cup 2 so that the levels in the two cups are equal.
6. Fill Cup 2 to the 200-mL mark with tap water from the bottle. Mix well.
7. Now pour half of the liquid from Cup 2 into Cup 3 so that the levels in the two cups are equal.
8. Fill Cup 3 with water to the 200-mL line with tap water. Mix.
9. Repeat this procedure, using Cups 4 through 9, until all nine cups have been filled. Note: Cup 9 will be completely full of solution.
10. To Cup 10 add 200 mL of tap water.
11. Record the dilutions in a data table like the one on page A-14.

Part Two

1. Each person in your group should take a clean, small tasting cup. This will be your tasting cup, so do not mix it up with the others. Beginning with the tap water in Cup 10, pour a small amount of water (about 15 mL, or enough to half fill the tasting cup) into your small tasting cup. This will represent the control—a solution that contains no drink mix. Look at the sample, smell it, and then take a taste. Record your observations. Empty your cup after each taste.
2. Pour a small amount from Cup 9 into each person's tasting cup.
3. Again, look at the sample, smell it, and then take a taste of the solution. Do not tell your group whether you can see, taste, or smell anything. Record your observations in your data table.
4. Repeat the process for Cups 8 through 1, in that order. For each cup, record whether you are able to see, smell, and taste the drink mix in each cup. Results of the tests should be kept private until all 10 solutions have been tested. Take care not to let your partners know through your body language or facial expressions whether you can detect the drink mix.
5. Circle on your data table each instance where you were first able to detect the drink mix (for example, circle the box under "Appearance" where you were able to see the mix).

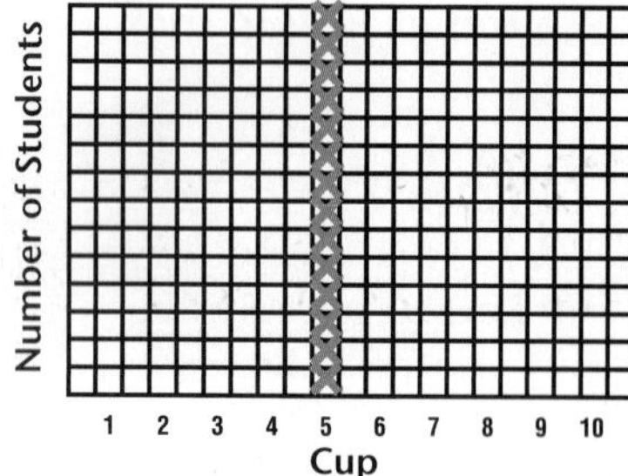

If everyone first saw the drink mix in Cup 5, your bar graph would look like this.

6. Record on Transparency 2, "Grids for Sensory Threshold Bar Graphs," the cups in which you were first able to see, smell, and taste the drink mix. Remember, you started with the very dilute drink mix in Cup 9, so the first cup in which you detected the drink mix was the highest, not the lowest, cup number.
7. Share your results with the other members of your group of four.
8. Complete the Analysis section.
9. Clean up as directed by your teacher.

Sensory Thresholds

Cup	Dilution	Appearance?	Smell?	Taste?
10				
9				
8				
7				
6				
5				
4				
3				
2	1/2			
1	1			

ANALYSIS

1. Use the bar graph sheet provided by your teacher or a sheet of graph paper to make a bar graph for each of the three senses. Record the information from the class results on your graph. Use a different color for each test.

2. According to the graph, in what cup were the most students first able to:

   a. Detect the color of the drink mix?

   b. Detect the smell of the drink mix?

   c. Detect the taste of the drink mix?

   These represent the most common values (modes) for vision, taste, and smell thresholds.

3. From your investigation, which sense is best able to detect the drink mix at low concentrations?

4. How did your personal results compare to those of the other members of your group?

5. Give two reasons that could explain why different people reported first seeing, tasting, or smelling the drink mix in different cups.

6. Describe an experiment you could do to check one reason you suggested for people sensing the drink mix at different concentrations. Include a statement of your problem and how you would do the experiment.

7. Based on the class discussion:

   a. Define the term threshold.

   b. Give an example of a threshold from this activity.

   c. Give an example of a threshold for a substance you use.

   d. Represent the concept of threshold. Use a labeled drawing and written description to express your ideas.

Parts per Million

In the last activity, you found that there were concentrations of powdered drink mix that you couldn't see, smell, or taste, but you didn't know exactly what the concentrations were. In this activity, you will find a way to describe the amount of food coloring in a solution if the amount is very, very small.

CHALLENGE

Use parts per million to describe the concentration of a solution. You will also be learning to perform a serial dilution, a procedure you will use many times in the activities.

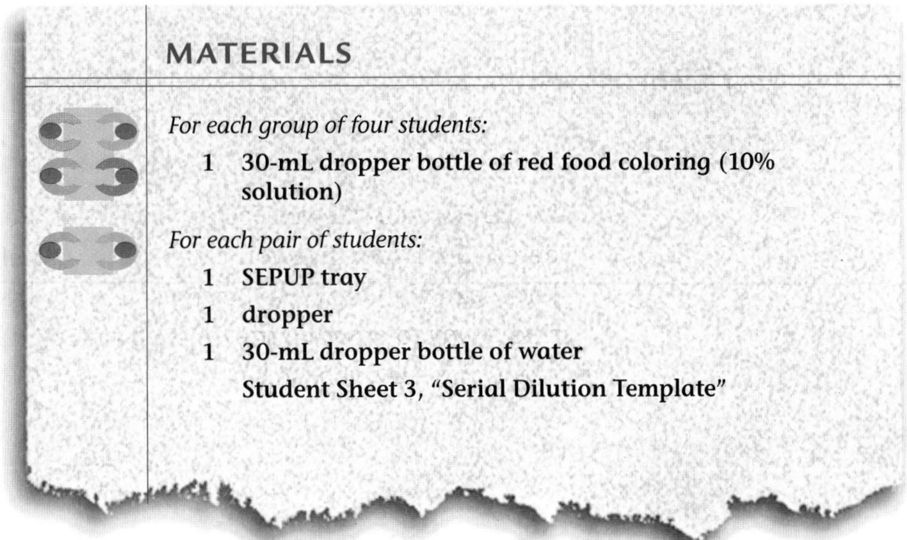

MATERIALS

For each group of four students:

1 **30-mL dropper bottle of red food coloring (10% solution)**

For each pair of students:

1 **SEPUP tray**

1 **dropper**

1 **30-mL dropper bottle of water**

Student Sheet 3, "Serial Dilution Template"

PROCEDURE

1. Place the Student Sheet 3, "Serial Dilution Template," under the SEPUP tray.
2. Put 10 drops of 10% red food coloring into small Cup 1 and put one drop into small Cup 2, on the lower level of your SEPUP tray.

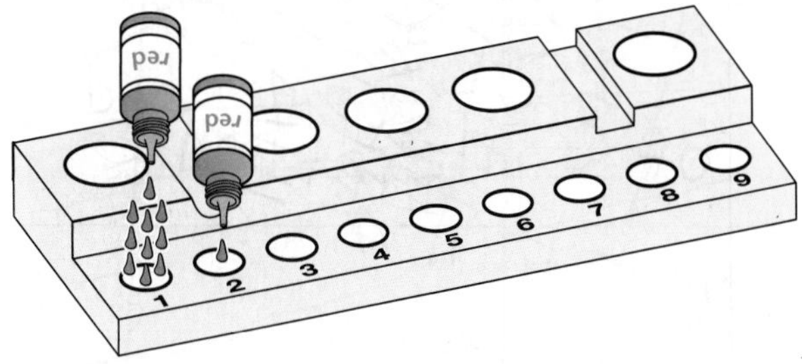

3. To small Cup 2, add 9 drops of water. Mix the solution by drawing it up into the dropper. Then gently squeeze the bulb until the dropper is empty, carefully putting the liquid back into Cup 2.

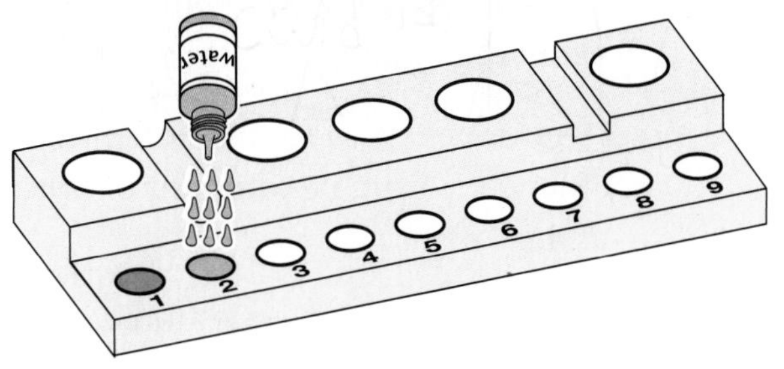

4. Using the dropper, transfer one drop of the solution in Cup 2 to Cup 3. Return any excess to Cup 2.

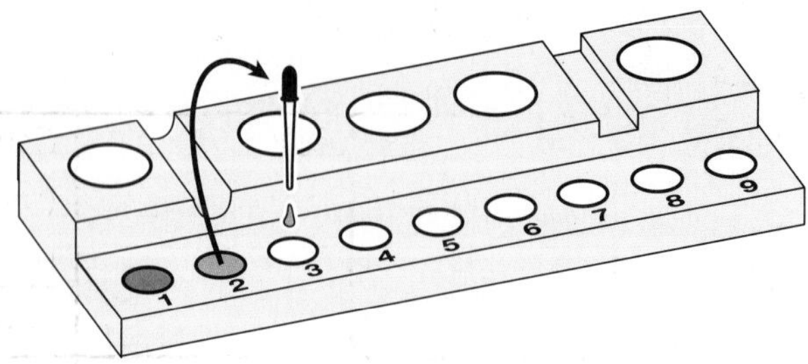

5. Add 9 drops of water to Cup 3. Use the dropper to mix the solution in Cup 3, and transfer one drop to Cup 4. Return any excess to Cup 3.

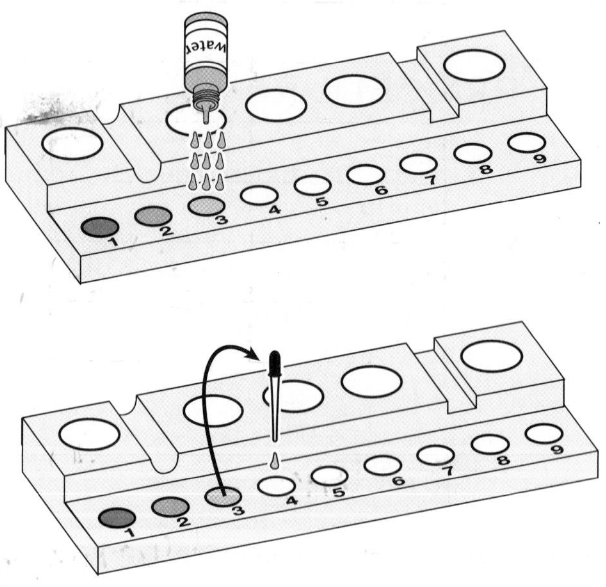

6. Add 9 drops of water to Cup 4. Mix. Transfer one drop to Cup 5. Add 9 drops of water to Cup 5. Mix.

7. Continue the process through Cup 9, each time taking a drop of the solution from the previous cup and adding 9 drops of water.

8. Record the color of the solution in each cup in a data table in your science notebook. A sample data table is shown on page A-19.

9. Determine the concentration of the solution for each cup as a part of food coloring per amount of solution, and record it in your data table.

10. Answer the Analysis questions.

Serial Dilution

Cup	Color	Concentration (parts of dye per parts of solution)	
		parts per ___	%
1		1 part in 10	10%
2		1 part in ___	
3		1 part in ___	
4			
5			
6			
7			
8			
9			

ANALYSIS

1. Which is more dilute, Cup 1 or Cup 2? How do you know this?
2. If Cup 1 has a concentration of one part in 10, and Cup 2 has 1/10 the concentration of Cup 1, what is the concentration of Cup 2?
3. Which cup has a concentration of one part per million?
4. What is the number of the cup in which the solution first appeared colorless? What is the concentration in parts of food coloring per parts of solution in this cup? (Express the answer for concentration as one part per ____.)
5. What are the possible reasons for student differences in reporting the cup in which the solution first appeared colorless? (**Hint:** Consider the idea of threshold.)
6. Do you think that any of the food coloring is present in this cup of diluted solution even though it appears colorless? Explain.

7. Explain how you could do an experiment to prove that there is actually some red food coloring in this cup.

8. How would you explain what a million is to a young child?

9. If you change the solution of food coloring in Cup 1 from one part in 10 (10%) to five parts in 10 (50%), what would the concentration of the food coloring be in Cup 6?

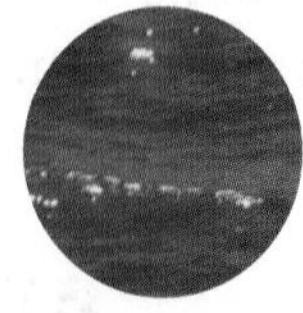

Some Interesting Comparisons

Now that you've spent some time trying to understand or picture just what one in a million (or even one in a billion or one in a trillion) means, here is a short list of comparisons.

Can you think of a one in a million comparison yourself?

- One part per million is one second in 12 days of your life.
- One part per billion is one second in 32 years of your life.
- One part per million is one penny out of $10,000.
- One part per billion is one penny out of $10,000,000.
- One part per million is one pinch of salt on 20 pounds of potato chips.
- One part per billion is one pinch of salt on 10 tons of potato chips.
- One part per million is one inch out of a journey of 16 miles.
- One part per billion is one inch out of a journey of 16,000 miles.
- One part per million is approximately one bad apple in 2,000 barrels.
- One part per billion is approximately one bad apple in 2,000,000 barrels.
- One part per billion is one square foot in 36 square miles.
- One part per trillion is a postage stamp in an area the size of New York City.

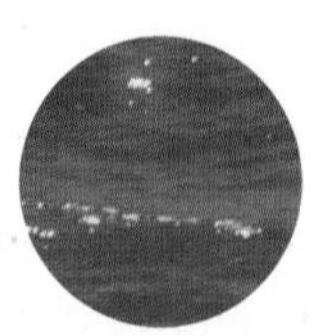

The Cholera Story

Picture yourself in London in 1832. What do you think life would be like? How would you dress? What kind of food would you eat? What would the air be like? What kind of house would you live in?

CHALLENGE

Cholera (CALL-er-ah) is a disease that is spread by a bacterium in water or through person-to-person contact. Place yourself in London in 1832 and imagine what it would be like if a member of your family were struck with cholera.

A tavern in London, named for John Snow, a London doctor who worked to help control the spread of cholera

Dr. William Brooke O'Shaughnessy was one of the first doctors to investigate the causes of cholera. He wrote the following observations in 1832:

> *Wanting to acquaint myself with the celebrated cholera, I traveled down to (London) from Edinburgh, prepared yet unprepared, dear sirs. I saw a face, a girl I never can forget, even were I to live beyond man's natural age.*
>
> *The girl lay . . . in a low-ceilinged room. I bent to examine her. The color of her skin—a silver blue, lead colored, ghastly tint; eyes sunk deep into deep sockets as though driven back or counter-sunk like nails, her eyelids black, mouth squared as if to bracket death; fingers bent, inky in their hue. Pulse all but gone at the wrist.*

This is another description of cholera:

> *It (is) not easy for survivors to forget a cholera epidemic. . . . The onset of cholera is marked by diarrhea, acute spasmodic vomiting, and painful cramps. Consequent dehydration (the victim can lose up to 5 gallons of liquid in 24 hours), often accompanied by cyanosis [the body turns blue], gives the sufferer a characteristic and disquieting appearance: his face blue and pinched, his extremities cold and darkened, the skin of his hands and feet drawn and puckered. . . . Death may intervene within a day, sometimes within a few hours of the appearance of the first symptoms. And these symptoms appear with little or no warning.*

From Charles E. Rosenberg, *The Cholera Years: The United States in 1832, 1849, and 1866.* Chicago: University of Chicago Press, 1962.

Cholera Deaths

In 1849, another outbreak of cholera killed over 500 people—rich and poor, young and old—in South London. John Snow, a medical doctor in England, had an idea. He thought that if he checked the city's death records and mapped exactly where people were living when they died, he might find some clues about what was causing the disease.

CHALLENGE

Examine the list of deaths from cholera in London in 1849 and plot their location on the map. See if there is a pattern that could explain how the disease spreads.

PROCEDURE

1. With your partner, use the following listing of cholera deaths to plot the locations of the victims' homes on the London street map that your teacher provides. (You'll need to tape the two pieces together to make one larger map.)
2. Use a colored marker to put a small dot at the approximate address for each death.
3. If there is more than one death at the same location, put the other dots as close as possible to each other. The grid location number will help you find the street addresses.

Figure 1: Deaths from Cholera in London in 1849						
Date	**Name**	**Age**	**Sex**	**Occupation**	**Address**	**Grid**
13 Feb	Anne Kelly	3	F	child	156 Broad St., between Marshall & Little Windmill Streets	E-5
23 Feb	Edwin Drummond	48	M	steeplejack	54 Little Windmill St., between Broad & Silver Sts	E-5
18 Mar	Patty Orford	23	F	seamstress	160 Broad St., near corner of Little Windmill St.	E-5
20 Mar	Sue Burton	22	F	seamstress	16 Queen St., near the corner of Little Windmill St.	H-3
27 Mar	Patrick Kelly	39	M	banker	156 Broad St., between Marshall & Little Windmill Streets	E-5
28 Mar	John Kelly	8	M	child	156 Broad St., between Marshall & Little Windmill Streets	E-5
3 Apr	Mary Thornley	45	F	governess	300 Marshall St., between Broad & Silver Streets	E-6
9 Apr	Thomas Topham, Jr.	19	M	butcher	8 New St., across from the brewery	E-4
9 Apr	William O'Toole	41	M	indigent	Poland Street Work House	D-6
13 Apr	Margaret Kelly	37	F	housewife	156 Broad St., between Marshall & Little Windmill Streets	E-5
21 Apr	Richard Raleigh	13	M	student	173 Broad St., between Poland & Marshall Streets	D-5
24 Apr	Katherine Nelson	1	F	child	426 Wardour St., next to the Brewery Yard	D-3
25 Apr	Russ Rufer	30	M	steeplejack	54 Little Windmill St., between Broad & Silver Sts.	E-5
29 Apr	Sarah Kelly	3	F	child	156 Broad St., between Marshall & Little Windmill Streets	E-5
1 May	Sir John Page	55	M	magistrate	255 Broad St., between Berwick & Poland Streets	D-4
2 May	Ann Nelson	19	F	housewife	426 Wardour St., next to the Brewery Yard	D-3
3 May	Agatha Summerhill	26	F	writer	174 Broad St., between New & Little Windmill Sts.	E-5
11 May	Barney Brownbill	31	M	indigent	Poland Street Work House	C-5
11 May	Rose Thornley	53	F	maid	300 Marshall St., between Broad & Silver Streets	E-6
17 May	Winnifred Topham	17	F	factory worker	2 Peter St., at the end	F-4
21 May	Thomas Topham	38	M	butcher	2 Peter St., at the end	F-4
22 May	Winston Page	49	M	doctor	1000 Regent St., near the corner of Hanover Street	D-9
27 May	Neville West	6	M	child	19 Golden Square	G-6
27 May	Beatrice Braxley	23	F	housewife	253 Broad St., between Berwick & Poland Streets	D-4
27 May	Eleanor Raleigh	12	F	student	173 Broad St., between Poland & Marshall Streets	D-5

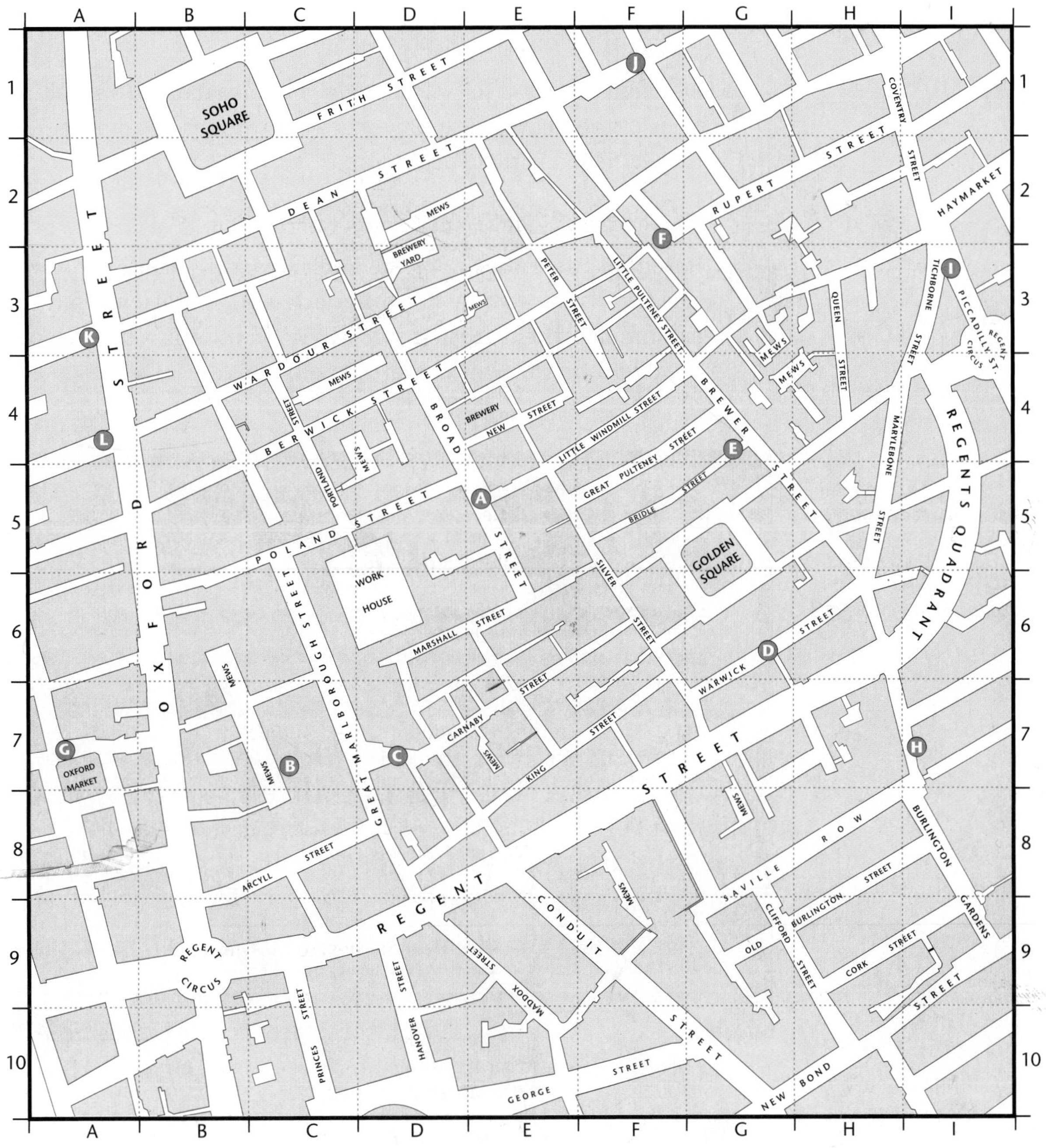

Map of London in 1854. Letters indicate major city water pumps.

ANALYSIS

1. Describe what you see on the map you have marked with the locations of the deaths. Are they scattered throughout the city, or are they bunched in a particular area?

2. Do you see any clues about the cause of the disease?

3. Based on the evidence of the cholera death locations shown on the map, state two or three **hypotheses**, or reasons that might explain how the disease is spread. (A hypothesis is an idea or theory about how or why something happens.)

CHOLERA.

THE

DUDLEY BOARD OF HEALTH,

HEREBY GIVE NOTICE, THAT IN CONSEQUENCE OF THE

Church-yards at Dudley

Being so full, no one who has died of the CHOLERA will be permitted to be buried after *SUNDAY* next, (To-morrow) in either of the Burial Grounds of *St. Thomas's*, or *St. Edmund's*, in this Town.

All Persons who die from CHOLERA, must for the future be buried in the Church-yard at Netherton.

BOARD of HEALTH, DUDLEY.
September 1st, 1832.

W. MAURICE, PRINTER, HIGH STREET, DUDL

Cholera deaths were often so frequent there were few places to bury the victims.

5 John Snow and the Search for Evidence

Snow's Theory

Mapping the locations of deaths caused by cholera led Dr. Snow to propose that cholera was being carried in the water.

Dr. John Snow

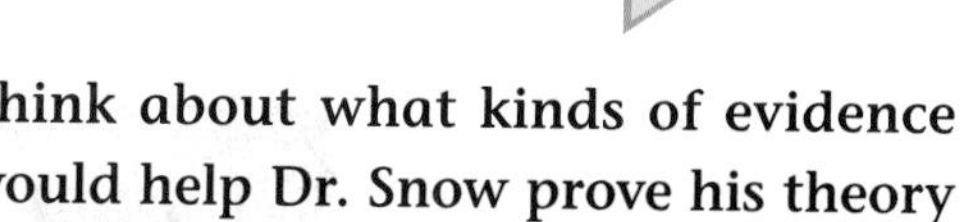

Think about what kinds of evidence would help Dr. Snow prove his theory about how cholera spreads.

DR. SNOW AND THE CHOLERA

Cholera causes severe diarrhea and vomiting. In India, cholera has been a health problem since 400 B.C., but it was not well known outside the Far East before 1800. In 1819, however, there was an epidemic in Europe and North America. Gradually, cholera almost disappeared, but it would sometimes reappear suddenly in one place or another. A person with cholera might die within a day, sometimes within only a few hours after the first symptoms appeared. People were confused about the cause of the disease because when some people got cholera, others living nearby would not. Bad air or piled-up trash were often considered to be the cause of cholera.

In 1849, a terrible outbreak of the disease killed over 500 people who lived within a few blocks of each other in London. John Snow, a medical doctor in England, drew a detailed map of the area, showing where each victim lived. Dr. Snow discovered that the deaths came mostly from houses located near a certain public water pump. Lots of people liked to drink from this pump because of the taste and clarity of its water. Some of the deaths were reported in houses farther away from the pump and did not immediately fit the pattern. Even so, after studying all of his data, Dr. Snow suggested that cholera was spread in the water supply by invisible bits of human waste from cholera victims.

FUN.—August 18, 1866.

DEATH'S DISPENSARY.

OPEN TO THE POOR, GRATIS, BY PERMISSION OF THE PARISH.

Cholera spread rapidly through London in contaminated drinking water.

Cholera was not just a London problem, as seen here in a cartoon from a New York newspaper from the 1800s.

Dr. Snow was concerned that his mapping of the 1849 cholera cases did not provide enough evidence to prove his theory. He began to carefully review public records covering the 20 years before the cholera outbreak. One thing he discovered was that since 1830, about 300,000 people in one area of South London had been served by just two water companies, the Lambeth Water Company and the Southwark and Vauxhall (S & V) Water Company. Originally, both companies drew their water from the Thames River in London. In 1840, however, the Lambeth Water Company changed and began to take its water from the Thames at a place 10 miles upstream. The other company, S & V, continued to get its water from inside London.

When cholera struck again in 1854, Dr. Snow reviewed the records for the area served by the two water companies. He asked his friend and colleague Dr. John Joseph Whiting to help him gather additional evidence concerning the water sources for the city of London. He wanted to provide more proof for his theory.

Searching for More Evidence

Dr. Snow and Dr. Whiting were two investigators who shared their data through writing. The S & V Water Company would not give them the information they wanted, so they had to work hard to collect it. Once they had the information, it had to be carefully organized before they could **infer** the cause of cholera. (Infer means to reach a conclusion on the basis of available evidence.)

GROUP CHALLENGE

The imaginary letters that follow are drawn from what is known about the correspondence between the two doctors at the time. As a group, decide if Dr. Snow's theory about the cause of cholera is consistent with the evidence provided in the letters between Dr. Snow and Dr. Whiting.

17 August 1854

John Joseph Whiting
47 Waterloo Road
London, England

Dear Dr. Whiting,

As you are aware, the dreaded cholera has once again come to our city. I have a great interest in showing the powerful influence which invisible bits of human waste in the drinking water have on the spread of cholera. The recent outbreak has given me the opportunity to test my theories on the grandest scale. The S & V Water Company will not give me data, so it will be necessary to collect it by going to each house. I feel the need of your help in this experiment. My current research includes the following results:

- *The area of the present outbreak is served by two water companies, S & V and Lambeth.*
- *Each of the companies supplies both rich and poor; large and small houses, and people of both sexes, all ages, and all occupations.*
- *Several years ago, the Lambeth Water Company moved its water intake pipe upstream from London's sewage-infested water. I am investigating whether this move is related to a decrease in the number of cholera cases. If so, it would support my idea that cholera is spread by human waste in the water.*

To provide evidence to support my theories, I need to learn which company supplied water to each house where a fatal attack of cholera occurred. Would you please collect the numbers of deaths that occurred at houses supplied by the S & V Water Company?

If you are agreeable, I will send you the numbers that I have already collected from houses served by the Lambeth Water Company.

Sincerely, your friend,

John Snow, M.D.

30 April 1855

Dear John,

The information that you requested was obtained with a good deal of trouble. Many hours were needed to collect the data, as I had to go door to door.

The analysis of my data should leave no doubt about the correctness of your hypothesis concerning the progress of cholera. I have combined our data and report it to you as follows:

- *S & V Water Company supplied 40,046 houses in which there were 1,263 deaths.*
- *Lambeth Water Company supplied 26,107 houses in which there were 98 deaths.*
- *The rest of London is served by other water companies. These companies supplied 256,423 houses in which there were 1,422 deaths.*

It has been a pleasure for me to be able to assist you. This experiment is of great importance to the community of London.

Sincerely, your friend,

John Joseph Whiting

INDIVIDUAL CHALLENGE

Make believe you are Dr. John Snow. Write a letter to the London Health Department describing your findings. Your letter should include the following:

- What you think is the cause of the spread of cholera
- All the data you have collected presented in an organized form
- Reasons the Health Department should believe your evidence
- What action you suggest the Health Department should take

6 Chlorination: How Much is Just Enough?

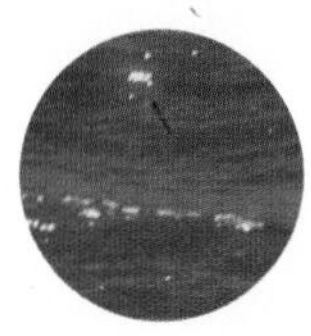

The Milwaukee Story

The risk of disease caused by biological contamination of drinking water is not just a thing of the past. Even modern cities have problems keeping water safe to use. In 1993, for example, the water supply of Milwaukee was threatened for several weeks. As a result, many people got sick and the entire city population was inconvenienced.

CHALLENGE

The article on the following page was based on those written at the time of the problem. Read it and be ready to discuss your ideas about what you think might have gone wrong with Milwaukee's water purification system.

Silver Oaks Beacon

April 23, 1993

Illness in Milwaukee caused by contaminated water

Public health officials in Milwaukee continue to gather information about an outbreak of intestinal illness that first struck in March. In early April, authorities announced that contamination of the municipal water supply may have caused the recent outbreak. They advised Milwaukee residents to boil all water for drinking, brushing teeth, or washing food.

Thousands of people in the Milwaukee area have suffered from unpleasant symptoms such as diarrhea, vomiting, and extremely painful stomach cramps. The cause of this outbreak appears to be cryptospiridosis, a disease caused by a parasite called cryptosporidium. This organism multiplies in the intestines of infected humans or animals (such as cattle) and is released in fecal wastes. It can then be transmitted in water.

The Milwaukee outbreak is the largest known outbreak of this disease. Most healthy people recover in about two weeks, but the disease can be fatal to people whose immune systems are weak. Very young children, the elderly, and people suffering from AIDS or other immune system conditions are most at risk.

Milwaukee residents have emptied the shelves of local stores of bottled water and intestinal remedies. "Get your Imodium here!," advertised a sign on one drugstore. Absenteeism rates have been high in city schools and offices.

Health officials think that recent heavy rains have caused run-off into Lake Michigan, the source of Milwaukee's water. This run-off may have washed wastes from local dairy farms or other sources into the lake. The failure of a system at one of Milwaukee's three water purification plants may have also contributed to the problem.

A Century of Chlorination

Chlorine gas can be deadly. The idea of drinking it in your tap water may seem crazy at first. In small enough amounts, however, chlorine does not make people sick, and most people can't even sense it (the concentration is below their sensory thresholds). In the United States, chlorine has been added to public drinking water for almost 100 years in order to reduce the risk of disease that could be caused by **microorganisms** (invisibly tiny bacteria, viruses, parasites, and other organisms) living in the water. Has it worked?

CHALLENGE

Read this short history of chlorine in the water supply. Think about what the next 100 years might be like if the practice were suddenly stopped.

Standing water must be treated for public consumption.

CHLORINE, THE WATER SUPPLY, AND PUBLIC HEALTH

The chlorination of public drinking water has been called one of the most significant public health practices of the twentieth century. Chlorine was first used in drinking water to remove odors long before its disinfectant powers were known. Because it also acts as a bleach, it removes color from water containing some organic (carbon-containing) chemicals. In the United States, it was first added to the public drinking water supply in 1908 in Jersey City, New Jersey.

As chlorination spread across North America, there was a dramatic decrease in the number of deaths caused by cholera and typhoid. Before chlorination was introduced to Toronto in 1910, the yearly death rate from typhoid fever was 44.2 deaths per 100,000 population. By 1928, the death rate had dropped to 0.9 deaths per 100,000 population (see chart).

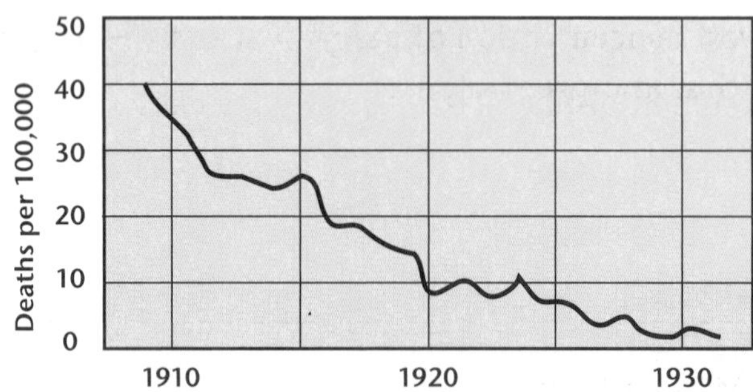

The incidence of these diseases had dropped to almost zero by the mid-twentieth century. The average life expectancy has increased from 49 years in 1900 to over 75 years today. This increase is due in part to an improved understanding of disease and rapid improvements in public hygiene, medical practices, and nutrition.

Taking water samples for later testing.

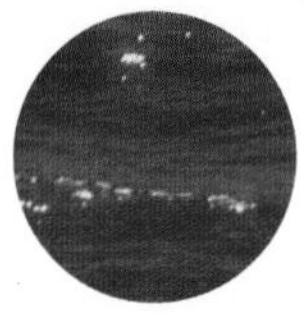

Determining an Effective Level of Chlorination

An effective way to kill microorganisms in water is to add chlorine. But chlorine can be harmful to humans if the concentration is too high. So, when treating water with chlorine, it is important to determine the lowest concentration needed to prevent biological contamination. In this lab, you will experiment with green algae, a harmless microorganism.

CHALLENGE

Determine the lowest concentration of chlorine measured in parts per million (ppm) that kills green algae.

MATERIALS

For each group of four students:

- **1 30-mL dropper bottle of household (5.25%) bleach solution**
- **1 plastic cup filled with about 100 mL of green algae water**

For each pair of students:

- **1 SEPUP tray**
- **1 30-mL dropper bottle of distilled water**
- **1 50-mL graduated cylinder**
- **1 dropper**
- **1 plastic cup half filled with tap water**
- **1 piece of white paper**
- **1 paper towel**

SAFETY NOTE: Do not bring chemicals into contact with your eyes or mouth. Wear safety eyewear as directed by your teacher. Avoid getting bleach on your clothing.

PROCEDURE

1. Set up your investigation report. Prepare a data table like the one on the next page.

2. With your partner, make a four-step serial dilution of the household (5.25%) bleach solution. This solution contains 13,000 ppm chlorine. Begin by putting 10 drops of the chlorine bleach into Cup 1, on the lower level of the SEPUP tray. Move one drop to Cup 2 and dilute it with 9 drops of the distilled water. Repeat this process until you get to Cup 4.

3. Carefully measure and pour 5 mL of the algae water into each of the five large cups (A–E) on the upper level of the SEPUP tray. Be sure to record the color of the algae water when you observe it against a white background.

4. Treat the water in Cup A with 5 drops of the concentrated chlorine solution from Cup 1. Treat Cup B with 5 drops of the first dilution from Cup 2. Cup C should be treated with 5 drops of the chlorine dilution in Cup 3, and Cup D with 5 drops from Cup 4.

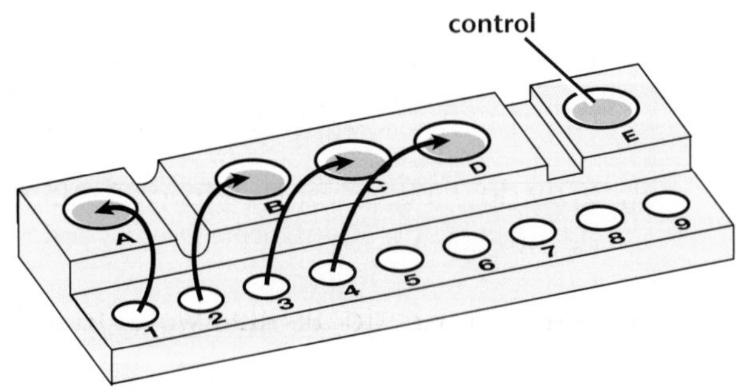

5. The algae water in Cup E is a control. Do not put any chlorine in it. Instead, put in 5 drops of the distilled water.

6. Carefully observe any color changes. Keep track of the time it takes for changes to occur.

7. Complete the Analysis section.

Water Treatment by Chlorination

Test Cups (5 mL algae)	Observations (before adding chlorine)	Treatment (5 drops)	Total Chlorine Concentration (ratio 1 to 25)	Observations (after adding chlorination)
A		25,000 ppm chlorine	1,000 ppm	
B				
D				
E				

ANALYSIS

1. Describe what happened in each of the five test cups during the time you observed them.
2. Based on your observations of the effect on algae, what concentration of chlorine would you recommend for water purification? Explain.
3. Is this a fair test for determining the appropriate concentration of chlorine to treat drinking water? Explain.
4. Why do we use algae instead of other microorganisms in our investigations?
5. Why did we add distilled water to the algae in Cup E?

Regulating Drugs: Another Public Health Issue

You have just finished examining some of the risks and benefits of chlorine. Do you ever think about the risks and benefits of prescription drugs? We usually think that prescription drugs are safe if we use them according to the label. It wasn't always that way.

CHALLENGE

Determine if the Food, Drug and Cosmetics Act (FDCA), which was passed in 1938, is still necessary today.

THE SULFANILAMIDE STORY

Before 1938, American drug companies could make and sell drugs without first testing them on animals or people. The drug companies didn't even have to get approval from the federal government. It was up to the government to prove that a drug was unsafe or mislabeled before it could be removed from the market. In 1938, the Food and Drug Administration (FDA) succeeded in obtaining legislation that required drugs to be cleared for safety before they could be sold.

In order to convince lawmakers of the need for laws regulating drugs, the FDA presented several stories documenting the ill effects of drugs

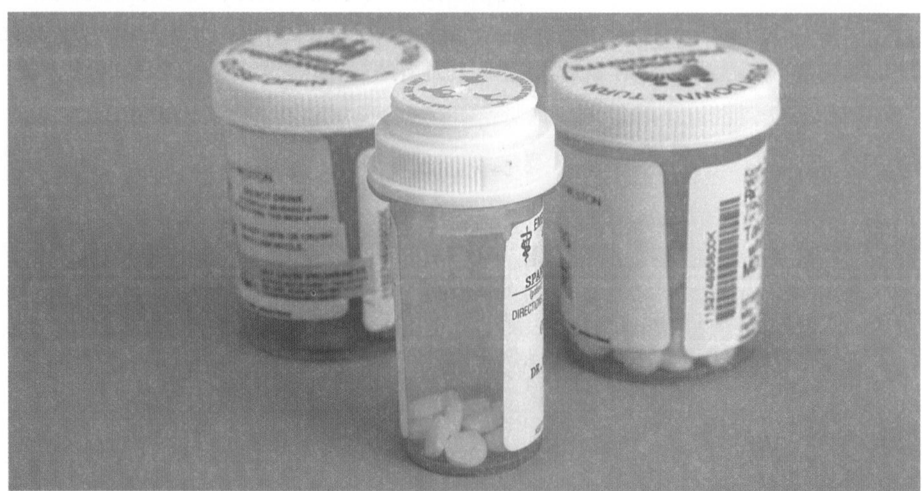

that were widely available at the time. The case that finally convinced legislators involved deaths produced by a liquid version of the drug sulfanilamide (sul-fa-NIL-a-mide).

This antibiotic was a new wonder drug at the time. It was one of the first antibiotics used to treat bacterial infections. Because it was hard for children to swallow the drug in pill form, the drug company found a liquid solvent that would dissolve the drug. It put the liquid form of the drug on the market without testing it. This liquid version of the drug killed 107 people, mainly children, before it was removed from the market.

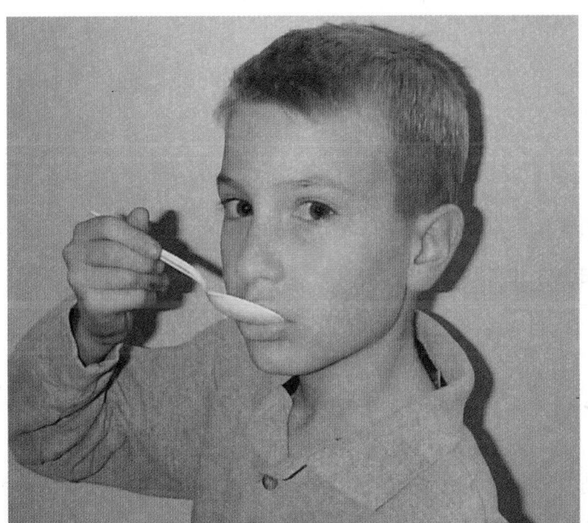

At the time, the drug was so new that no one knew whether it was the drug or the solvent that was toxic. Further tests showed that the deaths were caused by the solvent, diethylene glycol (die-ETH-el-een GLY-col), which is used in antifreeze. This tragedy led to the passage, in 1938, of the Food, Drug and Cosmetics Act. The act requires products to be tested, usually on animals, and approved for safety before they can be sold.

QUESTIONS

1. Describe in your science notebook what you think would have been an adequate pre-sale test of the safety of liquid sulfanilamide.

2. The Food, Drug and Cosmetics Act (FDCA) is over 65 years old. Is there still a need to keep this legislation on the books? Write a letter to the newspaper expressing your viewpoint in favor of or opposed to the FDCA.

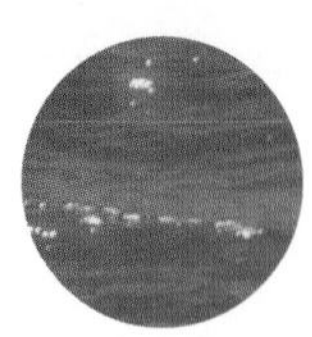

Testing for Chlorine in Water

Chlorine is commonly added to the drinking water supply to prevent bacteria and other microorganisms from growing in it. As you learned in Activity 6, the stronger the concentration of chlorine is, the more effective it is in killing microorganisms.

CHALLENGE

Test the three provided water samples to determine their chlorine concentrations.

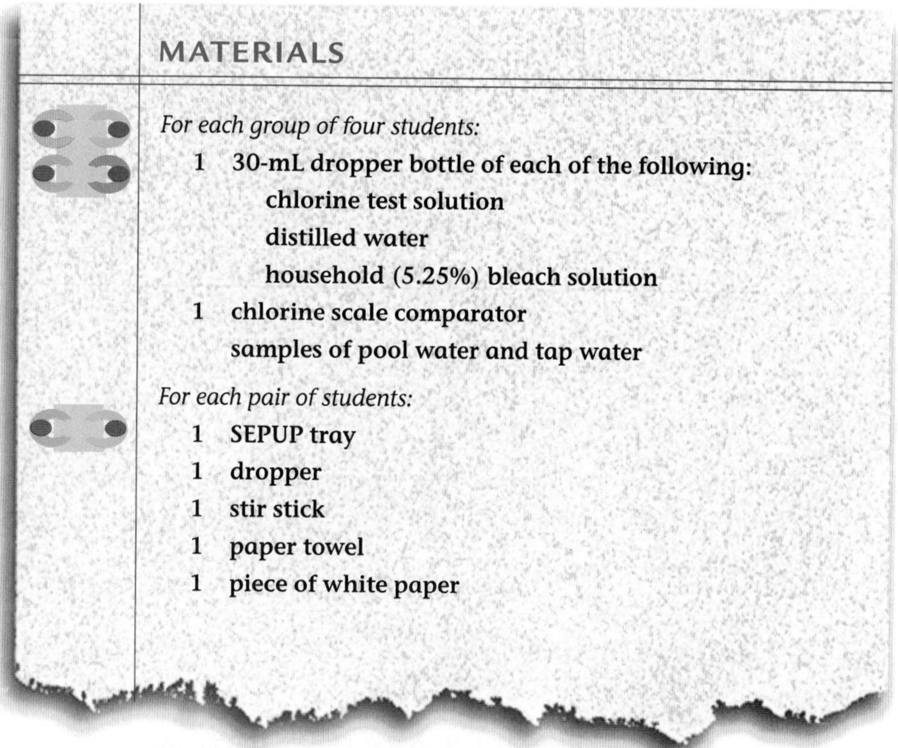

MATERIALS

For each group of four students:

1 30-mL dropper bottle of each of the following:
- chlorine test solution
- distilled water
- household (5.25%) bleach solution

1 chlorine scale comparator

samples of pool water and tap water

For each pair of students:

1 SEPUP tray

1 dropper

1 stir stick

1 paper towel

1 piece of white paper

SAFETY NOTE: **Do not taste or smell chemicals or bring them into contact with your eyes or mouth. Wear safety eyewear as directed by your teacher.**

PROCEDURE

Part One: Setting Up the Investigation

1. Assemble the SEPUP tray and other materials provided. Place a piece of white paper under the tray.
2. Place 10 drops of bleach solution into Cup 1, on the lower level of your tray.
3. Perform four step-by-step dilutions of the chlorine solution in Cups 2–5. Make each dilution a 1-in-10 (or 1/10) dilution. Use distilled water for the dilutions.
4. Add 10 drops of pool water to Cup 7.
5. Add 10 drops of tap water to Cup 8.
6. Add 10 drops of distilled water to Cup 9.

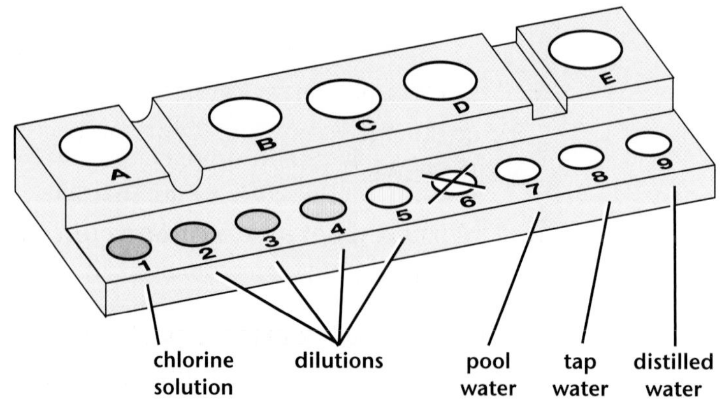

Part Two: Using the Comparator to Determine the Concentration of Chlorine

1. Decide how you will record your data.
2. Add one drop of chlorine test solution to Cups 1–5 and 7–9.
3. Hold a chlorine color comparator up to each cup. Compare the color in the cup to the comparator.
4. Estimate the concentration of chlorine in the cup in ppm. Record your estimate.

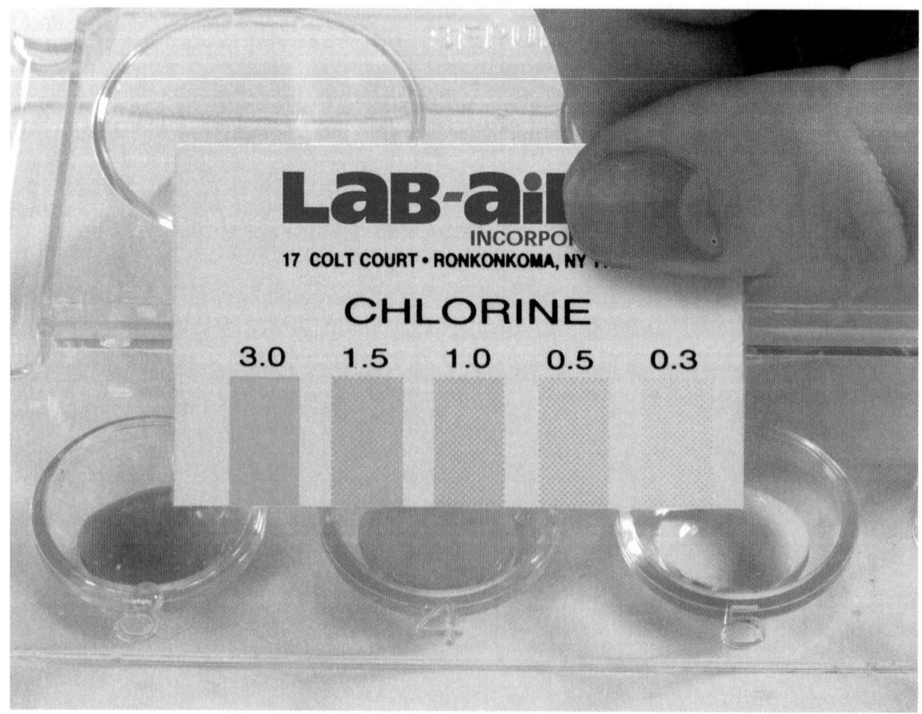

ANALYSIS

1. Did you notice any differences between the colors you found in the cups and those on the color comparator? What might cause those differences?

2. Did you compare each chlorine test with a control? If you did, what was the control?

3. The expected values for pool water are between 1.0 and 3.0 ppm chlorine. For tap water they are 0.2 to 0.5 ppm chlorine. How do your results compare with these values?

4. What concerns would you have if the concentration of chlorine in the pool water or tap water was too high? Too low?

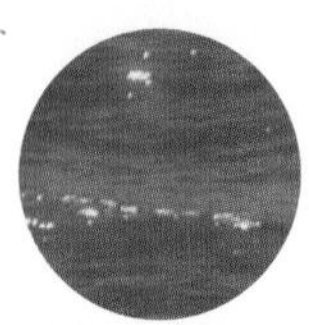

Tiny Cancer Risk in Chlorinated Water

You used chlorine to kill microorganisms and know that we use chlorine to purify drinking water. Is there a danger in using chlorine to purify drinking water?

Examine the facts presented in the following article and react to the possible danger of water chlorination to humans.

Silver Oaks Beacon

July 2, 1992

Chlorinated water linked to cancer risk

Chlorinated water may increase the risk of two forms of cancer, according to a study published yesterday in the *American Journal of Public Health*. The study was conducted by a collaborative group that included researchers from the Medical College of Wisconsin, Harvard University, the American College of Physicians and the Institute of Hygiene and Preventive Medicine in Naples, Italy.

The researchers used statistical methods to analyze the results of ten earlier studies. Their analysis showed a slight increase in the rates of two forms of cancer correlated with the amount of chlorine by-products in water. These by-products form when the chlorine added to water reacts with tiny amounts of other chemicals present in the water supply.

About three-fourths of the water supply in the United States is chlorinated to disinfect the water and prevent water-borne illnesses. The amount of chlorine added varies, depending on the degree of contaminants in the water. The amount of chlorine added to water now is often much less than was allowed in the 1970s, when the ten studies were done.

The research team found that the relative risk for bladder cancer for people drinking chlorinated water was 1.21, in comparison to a risk of 1.00 for people who drink non-chlorinated water. The relative risk for rectal cancer increased to 1.38. This means that the chances of getting bladder or rectal cancer increase by 21% and 38% for those who drink chlorinated water.

The risks of getting bladder or rectal cancer are relatively low, so the increased levels due to chlorination represent a low, but significant, risk of cancer. The authors of the article encourage a search for safer methods of disinfecting water.

QUESTION

Based on the information in the article, would you recommend chlorinating the drinking water in your community? Give your reasons.

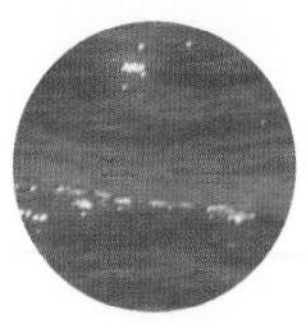

Cholera Epidemic in Peru

You have made a decision about chlorinating the water in your own community. The health officials in Peru in South America studied the evidence and made a decision. Read about their decision and see if this affects your thinking on the subject.

CHALLENGE

Read "Cholera Epidemic Sweeps Peru" and use all the evidence you have collected since the beginning of the course to consider the trade-offs involved in chlorinating the water supply.

Silver Oaks Beacon

November 29, 1991

Cholera epidemic sweeps Peru

A decision by Peruvian public health officials to halt the use of chlorine in water treatment efforts is being blamed for the cholera epidemic that is now sweeping Peru and many other countries in South and Central America. As of November 1991, more than 300,000 cases of cholera had been reported, mostly in Peru, and more than 3,500 lives had been claimed by the epidemic.

Pan-American Health Organization (PAHO) officials believe that the bacterium that causes cholera first arrived with a Chinese freighter that dumped contaminated waste water into the harbor of Lima, Peru. The organisms quickly spread to fish and shellfish and probably first infected humans in servings of raw fish called ceviche, a popular local dish. Once humans were infected, the disease spread rapidly through the local water supply. Cholera bacteria in an untreated water supply can thrive and infect many times as many people as might have otherwise been exposed by person-to-person contact.

According to an article in the international science magazine *Nature*, one explanation for the spread of cholera in Peru was that health officials in Peru might have misjudged the relative risks of water chlorination on one hand and microbial contamination on the other. Based on United States Environmental Protection Agency studies showing a possible increase in the risk of cancer due to the by-products formed by chlorinated water, officials from Peru halted chlorination of many of Lima's wells. Chlorine is a disinfectant that is capable of killing the bacterium that causes cholera, and is used to treat most city water supplies in the North America and Europe. However, chlorine also can react with organic substances present in most water supplies to produce several suspected human carcinogens, including chloroform.

Local government officials say that although Peru has good water filtration technology and pumps safe water into the drinking water system, old pipes and unchlorinated open wells may have allowed the cholera bacteria to enter the water supply after filtration. PAHO officials note that if chlorine were present, the cholera bacteria would have been killed, regardless of their point of entry into the system.

Ceviche, a popular South American dish

QUESTION

Does the information in this article change your opinion about chlorination of water? Explain.

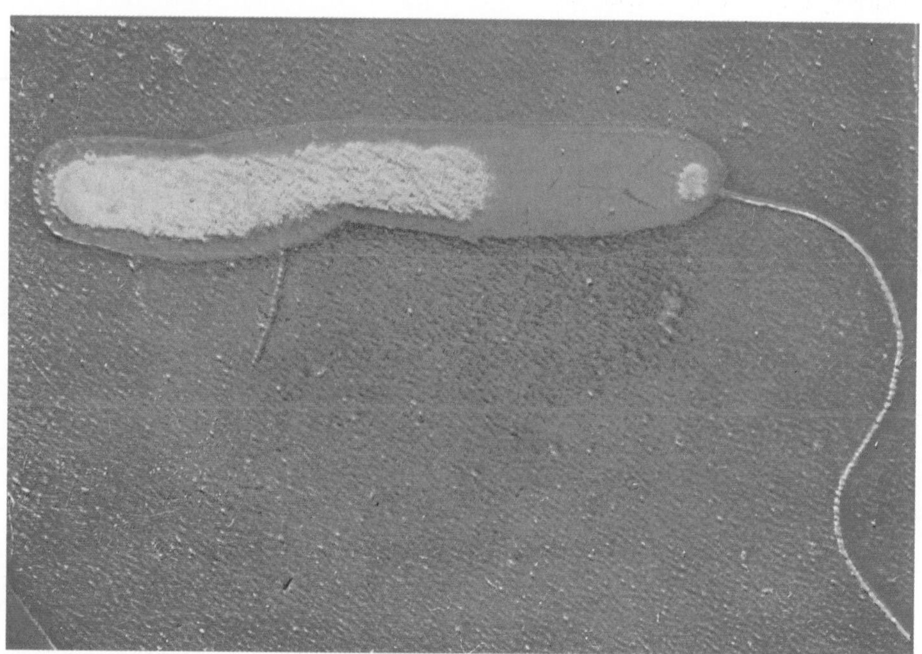

Cholera vibrio, *the bacteria that causes cholera*

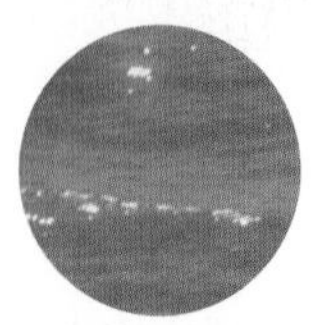

Chlorination Balance Sheet

The previous articles have given you information about the benefits and risks of chlorination. Assume these articles recently appeared in your local newspaper. The information in the articles has caused some community members to express concern about risks of cancer and waterborne diseases in the drinking water supply.

CHALLENGE

Use evidence to weigh the trade-offs involved in water chlorination.

You are a public health official who works in the Water Department. Your supervisor has asked you to respond to the public's concern about water chlorination at the next City Council meeting.

Prepare a written response explaining the issues that the newspaper articles raise. Be sure to discuss the advantages and disadvantages of chlorinating drinking water in your response. Then, explain your recommendation about whether the water in your town should be chlorinated.

8 Acids, Bases, and Indicators

The federal government establishes standards for water quality (see chart, page A-53). Water quality is affected by the chemicals dissolved in it. In the next few activities, you will learn an important way of classifying chemicals—as acids or bases. You will also learn techniques that enable you to measure how acidic or basic a substance is.

The pH of a solution is a value that is used to express how acidic or basic it is. The pH scale ranges from 0 (strongly acidic) through 7 (neutral) to 14 (strongly basic).

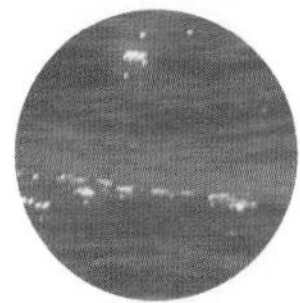

Acid Waste: An Environmental Issue

If the pH of water in rivers, streams, and lakes changes too much, that is if it becomes too high or too low, many animals and plants cannot survive in it. Even a small change from normal can harm some aquatic life and possibly affect the living things in the water with results that are hard to predict.

CHALLENGE

Read and analyze the following newspaper article. As you read it, think about what evidence and information you need to understand the issue described in the article.

Silver Oaks Beacon

City Officials Meet with C-Chip

Recent protests by citizens may hold up the construction of a computer chip manufacturing plant in Endicott City. The C-Chip Company has obtained a permit to build the plant in West Endicott, near the Endicott River.

Protesters say that the plant will dump acids in the river, and that will harm wildlife in the region. According to Jim Gallagher, representative of the West Endicott Coalition for the Urban Environment, the wastes released by the plant may "throw off the natural pH balance or the amount of salt in the river downstream from the plant. This could damage ecosystems in the river and the marsh where the river feeds into Great Lake."

The computer chip manufacturing process uses strong acid to produce the tiny electrical circuits on the chips. Vera Jackson, Environmental Consultant for C-Chip, stated, "The acid waste will be diluted and neutralized before it is released to the river. This neutral waste will have little or no impact on the diversity of wildlife in the river."

City officials are meeting with officials from C-Chip to discuss the impact of the proposed plant on the Endicott River.

QUESTIONS

1. Write down any words from the article that you don't understand.
2. List questions you would like answered or information you would like to have before making a decision about the issue described in the article.

National Primary Drinking Water Standards (Updated 1/02)[1]
U.S. Environmental Protection Agency

	MCL[2] (ppm)		MCL[2] (ppm)
1. Microorganisms			
Cyrptosporidium	99% removal	Total coliform	5%[3]
Giardia	99.9% removal	Turbidity (NTU)	1
2. Disinfection by-products			
Chlorine (as Cl_2)	4.0	Trihalomethanes	0.080
Haloacetic acids	0.060		
3. Inorganic chemicals			
Arsenic	0.05	Fluoride	4.0
Cadmium	0.005	Mercury	0.002
Chromium, total	0.1	Nitrate	10
Copper	1.3	Selenium	0.05
Cyanide (as free cyanide)	0.2		
4. Organic chemicals			
Benzene	.005	Dioxin	.00000003
Carbon tetrachloride	.005	Glypophosphate	0.7
Chlordane	.002	Toluene	1
2,4 D	.07	Trichloroethylene	.005

[1] This is not a complete list and is provided for comparison purposes only. The complete federal water standards tables provide maximum levels for more than 50 substances.

[2] Maximum Contaminant Level (MCL) - The highest level of a contaminant that is allowed in drinking water. MCLs are enforceable standards, by law.

[3] No more than 5.0% samples testing positive in a month. There may not be any fecal coliforms or *E. coli*.

Developing a Definition for Acids and Bases

Some substances can be used to test for other substances. These test substances are called indicators because they indicate (or show) that another kind of substance is present. One indicator is called phenolphthalein (FEEN-ul-THA-leen). In this activity you will learn how to use phenolphthalein and other acid-base indicators to investigate a variety of liquids in order to identify whether they are acids or bases.

CHALLENGE

Use your observations about the liquids you investigate to develop a definition that you can use to identify acids and bases.

MATERIALS

For each group of four students:

- 1 pH scale or color chart
- samples of household substances
- 1 30-mL dropper bottle of each of the following:
  - 1% sodium hydroxide (NaOH) solution
  - 1% acetic acid (CH_3COOH) solution
  - Universal indicator solution
  - 0.1% phenolphthalein solution

For each pair of students:

- 1 SEPUP tray
- 1 stir stick
- 1 dropper
- 9 pieces (half strips) of pH paper
- 1 paper towel
- 1 30-mL dropper bottle filled with water

SAFETY NOTE: **Wear eye protection and keep the chemicals from touching your skin.**

PROCEDURE

1. Set up your investigation report. You will test nine liquids with three different indicators. Think about how to set up a data table to record the information from these tests. Your teacher may have some suggestions.

2. Using the diagram below, set up your tray as follows:

   a. Add 5 drops of water to Cups 1–3, on the lower level of your SEPUP tray.

   b. Add 5 drops of acetic acid to Cups 4–6.

   c. Add 5 drops of sodium hydroxide to Cups 7–9.

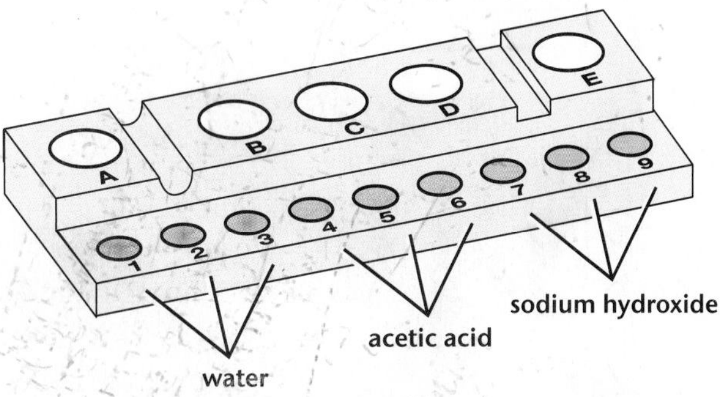

3. Test each liquid with three indicators as follows:

   a. Use one drop of phenolphthalein to test the liquids in Cups 1, 4, and 7.

   b. Use one drop of universal indicator to test the liquids in Cups 2, 5, and 8.

   c. Test the liquids in Cups 3, 6, and 9 with pH paper.

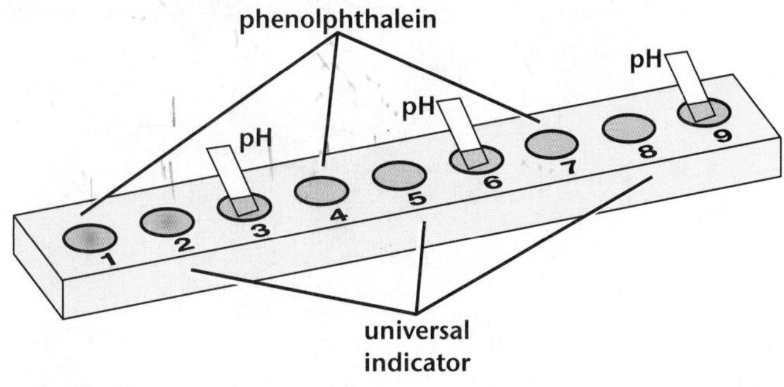

4. Using your data table, record the colors you observe. Rinse and dry your trays.

5. Your teacher will provide various household liquids for testing. Choose any three liquids for your first set of tests. Follow the same testing procedure that you used in Steps 2 and 3 to test each liquid with the three indicators. Record the colors you observe. Rinse and dry the tray.

6. Follow the same procedure to test another three household liquids. Record the colors you observe before cleaning the tray.

ANALYSIS

1. Use the results you observed to put the nine substances you tested into groups based on how they interact with the indicators.

2. Which seems to be the most useful indicator? Explain your answer.

Common household items may be acids or bases, or neutral substances.

9 Serial Dilutions of Acids and Bases

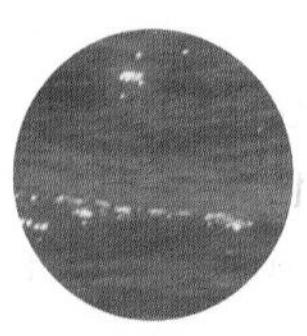

Investigating Acid and Base Dilutions

In Activity 8 you observed that there are indicators that can tell you whether a substance is an acid or base. You also learned that some indicators are useful over a wide range, while others can only be used over a more narrow range.

For example, phenolphthalein can only be used to detect bases; it cannot distinguish between neutral substances and acids. In contrast, universal indicator and pH paper are wide-range indicators; they provide information throughout the range from extreme acid to extreme base. You will now examine how you can use a standard scale to estimate the pH number of a solution.

Vinegar and many photographic chemicals are weak acids.

CHALLENGE

Set up a serial dilution to see how indicators behave as the concentration of acid and base changes. Look for patterns in the behavior of serial dilutions of acid and base when they interact with universal indicator.

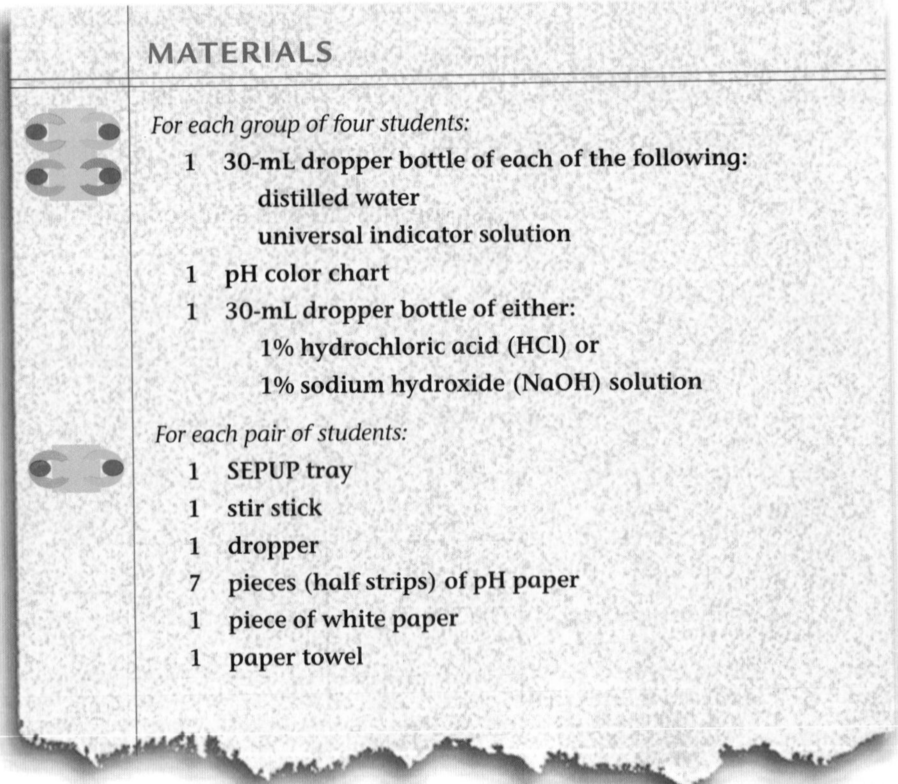

MATERIALS

For each group of four students:

1 30-mL dropper bottle of each of the following:
- distilled water
- universal indicator solution

1 pH color chart

1 30-mL dropper bottle of either:
- 1% hydrochloric acid (HCl) or
- 1% sodium hydroxide (NaOH) solution

For each pair of students:

1 SEPUP tray
1 stir stick
1 dropper
7 pieces (half strips) of pH paper
1 piece of white paper
1 paper towel

SAFETY NOTE: **Do not taste or touch the solutions. Use safety eyewear. Wash your hands after completing the activity.**

PROCEDURE

1. Set up your investigation report and data table so they are similar to the example shown below.
2. Put 10 drops of water in Cup 7, on the lower level of your SEPUP tray.
3. Perform a serial 1/10th dilution of the solution you are testing (A or B) in Cups 1–6. (If necessary, refer to Activity 3 to remind yourself about the dilution procedure.) Put water in one of the large cups for rinsing the dropper and stir stick between dilutions.
4. Test the pH of each solution with pH paper. Record the results.
5. Test the pH of each solution with universal indicator. Record the results.

Dilution of Solution ____

Cup	Dilution	Color and pH	
		pH paper	Universal Indicator
1	1/100		
2	1/1000		
3			
4			
5			
6			

6. Arrange the tray of acid and the tray of base so that the small cups are arranged, left to right, from most acid to most base, with the neutral cups in the middle. To do this, you will have to place the trays end to end. You may wish to have them overlap where they are neutral.

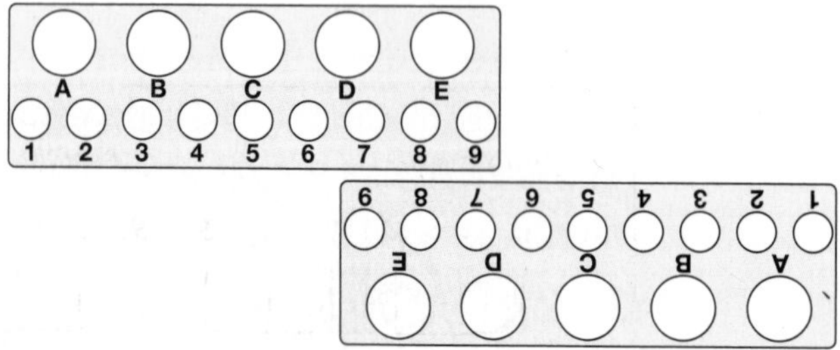

7. Complete the Analysis section.

ANALYSIS

Refer to the results of your group's investigation and the information in the following reading, "Acids, Bases, and pH" to answer the following questions.

1. Draw a diagram of your trays that you arranged from most acid (left side) to most base (right side).

   a. Label the cup that contains the known standard with pH 7, which is neutral.

   b. Draw an arrow that points from pH7 to the strong acid.

   c. Draw an arrow that points from pH7 to the strong base.

   d. Label each cup with the color and approximate pH of the solution.

2. How does the color change in your serial dilution relate to pH? Write a brief description that you could use to teach someone about this.

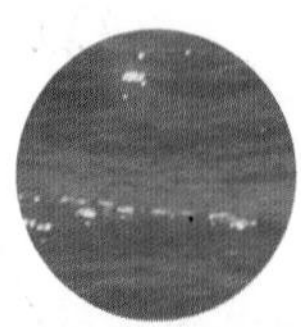

Acids, Bases, and pH

The pH of a solution is a value that expresses the solution's acidic or basic nature. A pH value may fall anywhere on a scale from zero (strong acid) to 14 (strong base). A value of 7 represents neutrality.

Strongly acidic Neutral Strongly basic

0 1 2 3 4 5 6 7 8 9 10 11 12 13 14

Bacteria

Plants (algae, rooted)

Carp, catfish, some insects

Bass, bluegill

Snails, clams,

Largest variety of animals (trout, mayfly nymphs, stonefly nymphs, caddisfly larvae)

Many aquatic plants and animals can live only within certain pH limits.

Relate the information in the reading to what you learned about pH in your serial dilutions of acid and base.

PH AND ITS MEASUREMENT

The measurement and control of pH is important in the manufacture of foods, paper products, and chemicals. In agriculture, it is necessary to maintain the correct soil pH for good yields of crops such as wheat, barley, corn, and other fruits and vegetables. Scientists also measure pH when they study acid rain and when they work to maintain water quality.

The federal government's standard for drinking water requires the water to have a pH between 6.5 and 8.5. If drinking water is too acidic or basic, it may be harmful to human health. It also may react chemically with the water pipes (which are usually made of metal) to produce byproducts that contaminate the drinking water. Many aquatic animals and plants cannot survive in pH levels below 6 or above 8.5.

Using indicators—substances that change color with a change in pH—is a way to measure pH. The color changes are the result of chemical reactions between the indicator and the acid or base. Acids turn pH paper red. Bases turn pH paper blue or violet. Acids and bases, and the salts produced by chemical reactions between them, are major groups of chemical substances. Originally, acids were recognized by their sour taste and because they could attack and dissolve some metals. Base solutions are usually slippery to the touch (think of bleach) and react with acids in water to form salts.

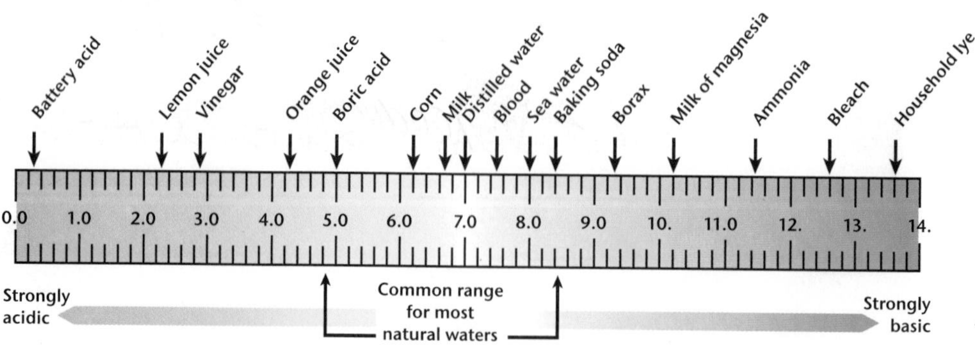

pH Scale

Acids and bases have important uses in the manufacture of many products. For example, in 2000, more than 5 million short tons of sulfuric acid were used by industry to make other chemicals and chemical products (a short ton is 2,000 lbs.). It is used in making fertilizer, refining petroleum, and in the industrial cleaning and processing of metals. Nitric acid is used in manufacturing explosives and dyes. The bases sodium hydroxide and potassium hydroxide are used in soap-making. Ammonia, another base, is used as a fertilizer and in the production of other fertilizers.

QUESTIONS

1. Why is the pH of water important to living things?
2. Explain how the dilutions in your trays can be used as a model for the pH scale shown on page A-62.

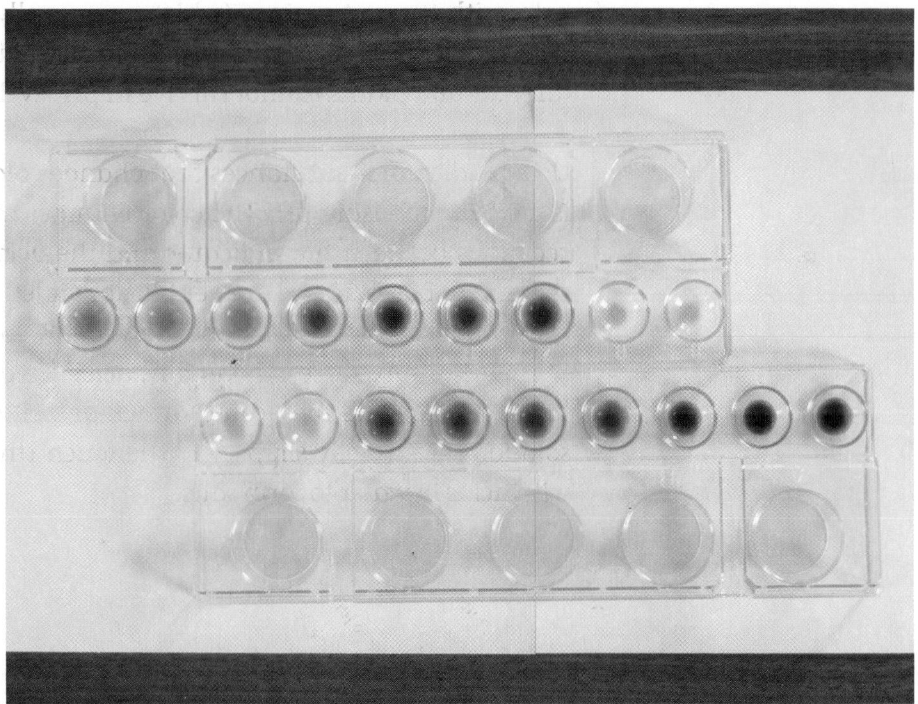

10 Acid-Bases Neutralization

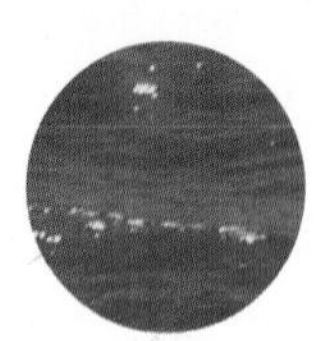

Mixing Acids and Bases

So far you have learned that indicators can be used to determine whether solutions are acids, bases, or neutral. You have also learned that indicators change colors, depending on the concentration of the acid or base solution. Finally, you learned that as acids and bases are diluted with water, they interact with the indicator more like water.

CHALLENGE

Begin to investigate what happens at the molecular level when solutions become neutral. Use your observations to help you develop a model for neutralization.

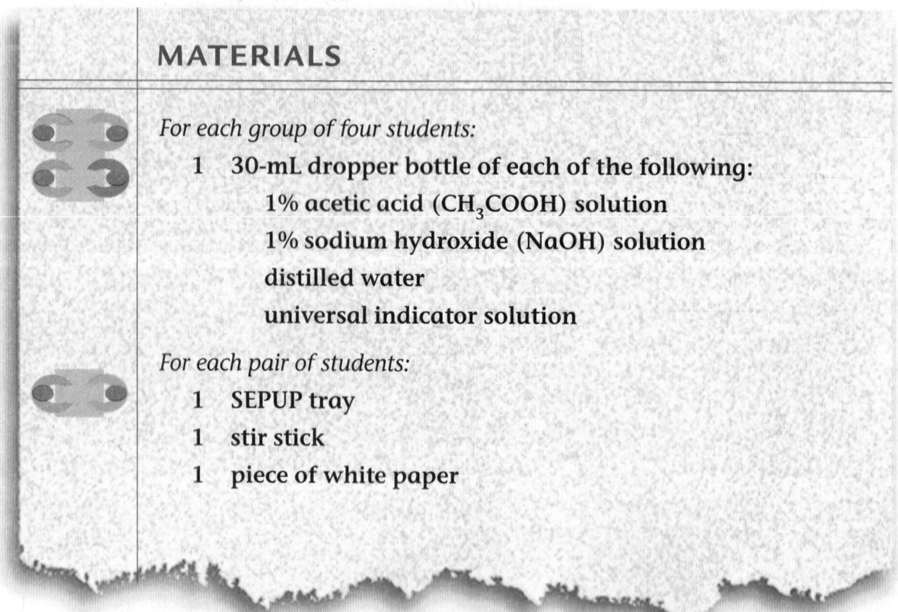

MATERIALS

For each group of four students:

1 **30-mL dropper bottle of each of the following:**
- **1% acetic acid (CH_3COOH) solution**
- **1% sodium hydroxide (NaOH) solution**
- **distilled water**
- **universal indicator solution**

For each pair of students:

1 **SEPUP tray**
1 **stir stick**
1 **piece of white paper**

SAFETY NOTE: **Wear safety eyewear. Avoid spilling the solutions on your skin. Be sure to wash your hands at the end of the activity.**

PROCEDURE

1. Set up your investigation report.
2. Place 10 drops of sodium hydroxide solution (base) in Cup 1, on the lower level of your SEPUP tray, and 10 drops of water in Cup 9.
3. Add one drop of universal indicator to Cups 1 and 9.
4. Add acetic acid one drop at a time, to the base in Cup 1. Stir after each drop. Record the colors you observe after each drop is added. Be sure to record the number of drops that turn the base the same color as the water in Cup 9. Stop when the mixture ~~turns red.~~ changes color.
5. Record on the class bar chart the number of drops of acid you needed to neutralize the base.
6. Place 10 drops of acetic acid in Cup 5.
7. Add one drop of universal indicator to Cup 5.
8. Add sodium hydroxide one drop at a time, to the acid in Cup 5. Stir after each drop. Record the colors you observe after each drop is added. Be sure to record the number of drops that turn the acid the same color as the water in Cup 9. Stop when the mixture turns blue.
9. Record on the class bar chart the number of drops of Solution 1 (the base) needed to neutralize the acid.
10. Follow your teacher's instructions for reporting your findings.

ANALYSIS

1. Which solution seems stronger, the acid or the base?
2. Explain your answer to Question 1.

A Model for Acid-Base Neutralization

You have observed that when you mix an acid and a base together in just the right amounts, the resulting solution is neutral. That is, the resulting solution shows neither acidic nor basic properties when tested with an indicator. How can we explain this observation?

Sometimes it is helpful to make a model to understand how something works. Many people do this, not just scientists. A model can take several forms. When an architect draws plans for a building, he or she may also make a small model of it to show people what it will be like. The architect's drawings and blueprints are also models of the building. A model can be a physical model, such as a miniature copy of a building, a diagram, such as an architect's blueprints, or even a mathematical equation.

CHALLENGE

Make another kind of model for the acid-base neutralization. Use your model to help you answer some questions about acid and base particles.

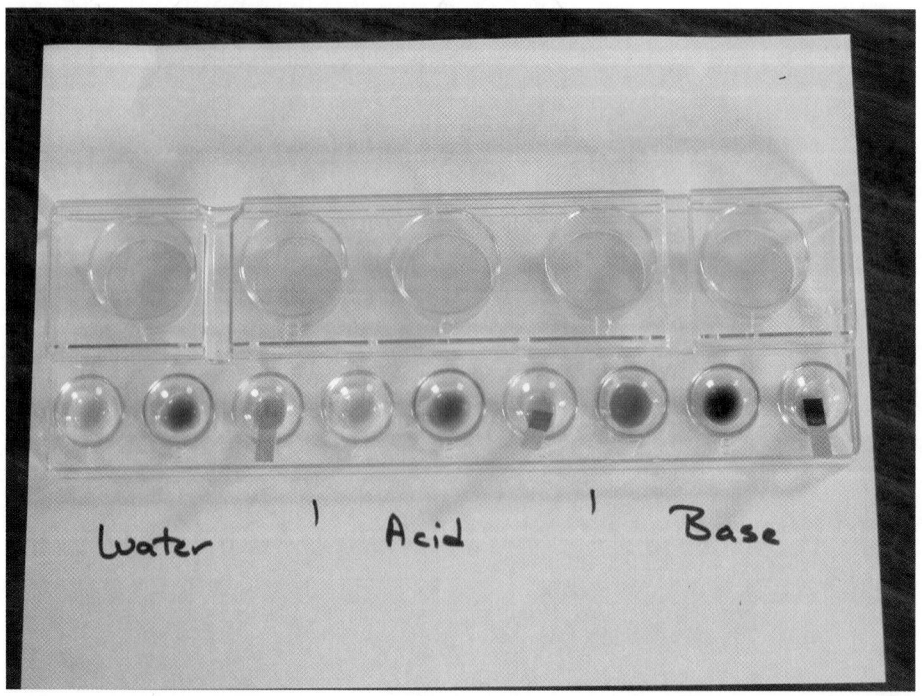

THE MODEL

In the following diagrams, each drop shape represents one drop of solution and each triangle represent a particle of acid or base. Suppose, as shown here, that each drop of acid contains the same number of particles as each drop of base. Then we can see that when one drop of base is added to one drop of acid, we have combined equal numbers of particles of acid and base.

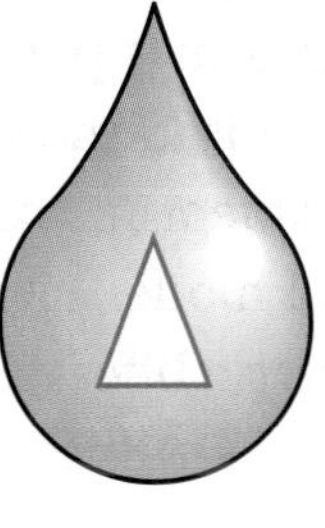

acid

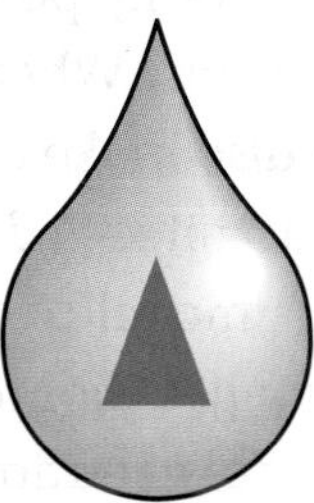

base

If, however, the acid is twice as concentrated as the base, then each drop of acid has twice as many particles as each drop of base. Therefore, 2 drops of base are needed to neutralize one drop of acid.

acid

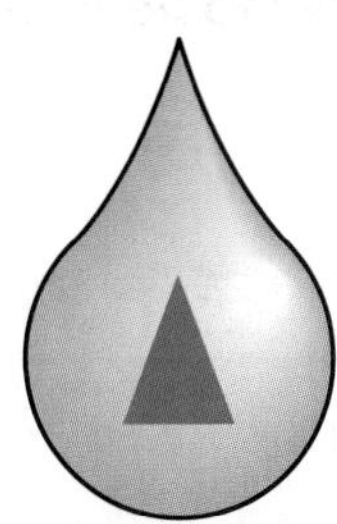

base

QUESTIONS

1. According to the bar chart of the class data, are there more particles of acid in a drop of 1% acetic acid solution or more particles of base in a drop of 1% sodium hydroxide solution?

2. Explain your answer to Question 1.

3. In your science notebook, illustrate drops of solution in which each drop of base contains twice as many particles as each drop of acid. Use △ for acid particles and ▲ for base particles.

4. According to the diagram in Question 3, how many drops of acid would be needed to neutralize 10 drops of base? Explain.

5. In your science notebook, illustrate drops of solution in which each drop of acid contains three times as many particles as each drop of base. Use △ for acid particles and ▲ for base particles.

6. In the example in Question 5, how many drops of base would be needed to neutralize 25 drops of acid?

7. In the example in Questions 5 and 6, how many liters of base would be needed to neutralize

   a. one liter of acid?

   b. 10 liters of acid?

   c. 20 liters of acid?

8. A student conducts an investigation to see which is more concentrated, a sample of acid or a sample of base. He finds that it takes 4 drops of acid to neutralize one drop of base. Which is more concentrated, the acid or the base? Explain your answer. You may wish to include a diagram.

11 Quantitative Analysis of Acid

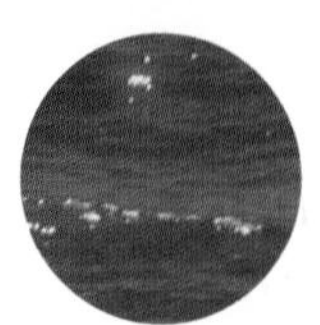

You have observed how universal indicator and pH paper can be used to give a rough estimate of the concentration of an acid or base solution. However, pH indicators cannot show the difference between 1% and 2% acid or base. When we need a more exact measure of the level of an acid or base in a solution, we need to use a different kind of test. This kind of test, which tells us about the amount, or quantity, of an acid or base, is called a quantitative test.

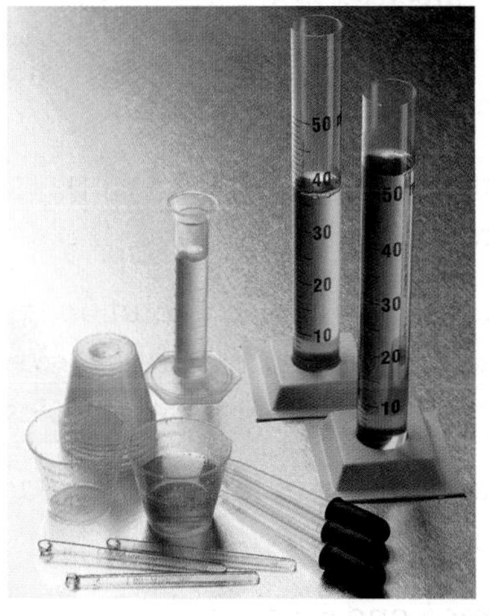

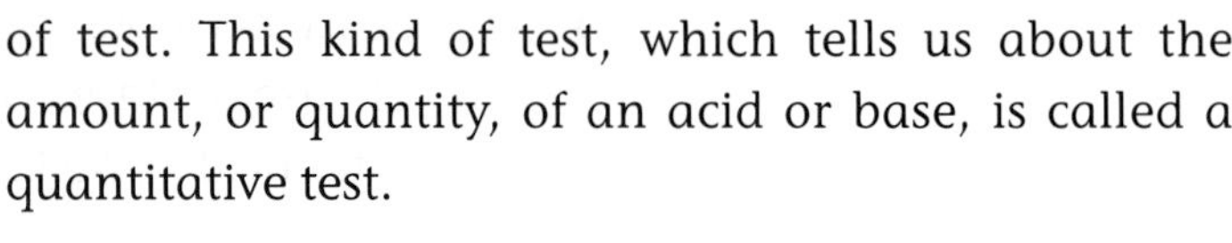

CHALLENGE

Your challenge is to quantitatively determine the concentration of an unknown acid sample. (To determine something quantitatively means to find out how it can be expressed in measured quantities.)

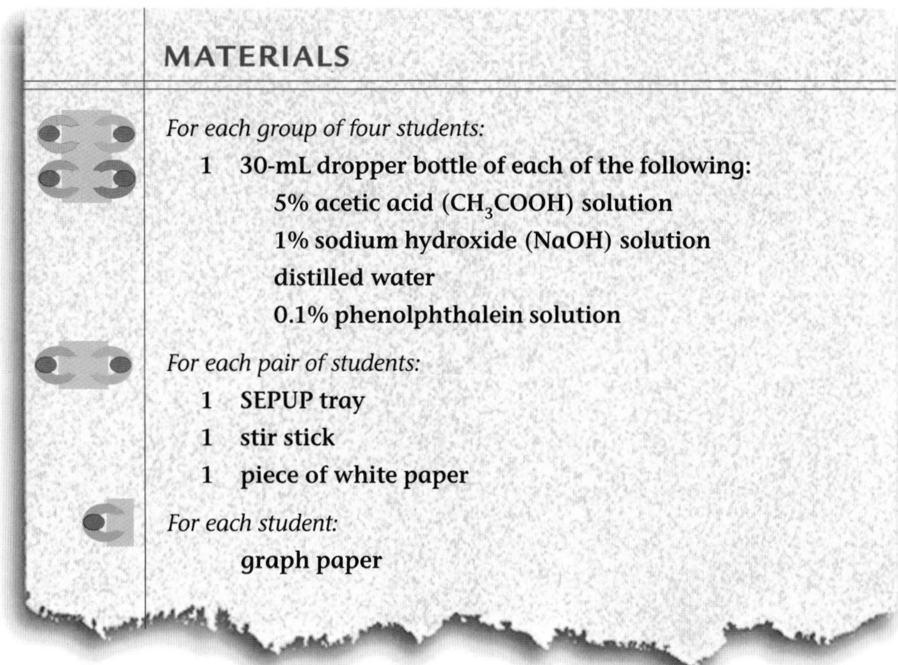

MATERIALS

For each group of four students:

1 **30-mL dropper bottle of each of the following:**
- **5% acetic acid (CH_3COOH) solution**
- **1% sodium hydroxide (NaOH) solution**
- **distilled water**
- **0.1% phenolphthalein solution**

For each pair of students:

1 **SEPUP tray**
1 **stir stick**
1 **piece of white paper**

For each student:

graph paper

SAFETY NOTE: Do not taste or touch the solutions. Follow your school policy regarding the use of safety eyewear. Wash your hands after you complete the activity.

PROCEDURE

1. Set up your investigation report form.
2. Add 2 drops of 5% acetic acid (CH_3COOH) to small Cups 1, 2, and 3, on the lower level of you SEPUP tray. Then add 8 drops of water to each of these cups.
3. Add 6 drops of 5% acetic acid (CH_3COOH) to small Cups 4, 5, and 6. Then add 4 drops of water to each of these cups.
4. Add 10 drops of 5% acetic acid (CH_3COOH) to small Cups 7, 8, and 9.

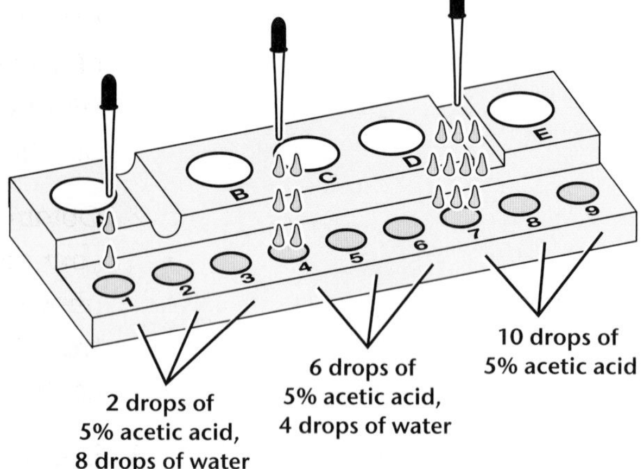

5. Calculate the final % of acid in each cup. Record the percentages you calculate on your investigation report.

   For example, in Cup 1, 2 drops of 5% acetic acid (CH_3COOH) were diluted with 8 drops of water. This means that 2 drops out of 10 drops, or 2/10, of 5% acetic acid was used:

   (2/10) × (5%) = 1%

6. For Cups 1–9, add one drop of 0.1% phenolphthalein solution to each cup.
7. One drop at a time, add 1% sodium hydroxide solution to Cup 1. Stir after each addition. Continue until the solution remains pink after stirring. It may be very pale pink or deep pink. Record the number of drops added.
8. Continue the procedure in Step 7 for each of the 9 cups. Be sure to record your results.

9. Calculate the average number of drops needed to neutralize the acid in Cups 1–3, Cups 4–6, and Cups 7–9. Record your calculations. Prepare a data table to summarize your results.

10. Prepare a line graph of the average number of drops needed to neutralize each concentration of acid. This graph is called a standard graph because it shows the results obtained with standard (known) concentrations of acid.

11. Do not clean up until you have completed Step 2 of the Analysis section.

ANALYSIS

1. Explain how you could use the standard graph and the procedure you just followed to find the concentration of an unknown sample of acetic acid. (**Hint:** How many drops of unknown would you need? What would you add to the unknown? How would this help you find out the unknown's concentration?)

2. Obtain some unknown acetic acid (Unknown X or Unknown Y) from your teacher. Carry out the procedure you described in Question 1 above. Use one of the large cups on the upper level of your SEPUP tray. Be sure to record your result. Explain how it tells you the concentration of the unknown acid.

3. Use your standard graph to predict how many drops of base would be needed in a similar experiment to neutralize 10 drops of 2.5% acetic acid or 10 drops of 4.5% acetic acid. Record your predictions. This procedure is called **interpolation** of data because you are determining the concentrations of solutions that fall *between* known points. (*Inter-* is a prefix that means "between.")

4. Use your standard graph to predict how many drops of base would be needed in a similar experiment to neutralize 10 drops of 6% acetic acid. Record your prediction. This procedure is called **extrapolation** of data because you are determining the concentrations of solutions that fall *outside* the known points. ("Extra-" is a prefix that means "outside.")

5. Which procedure (extrapolation or interpolation) would you have more confidence in? Explain your answer.

12 The "Used" Water Problem

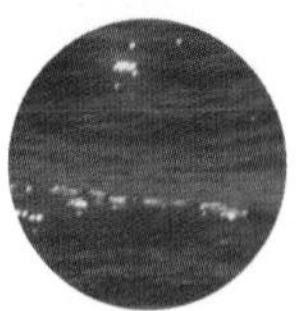

Imagine you work as the Safety Officer for the C-Chip plant that you read about in Activity 8. The plant uses acid solutions in a process that coats pieces of metal with chromium or gold. After the metal is coated, the acid is washed off. The water used to wash the acid off is called "used" water. It has an unknown amount of acid in it.

The C-Chip plant was built on the banks of a large river so that large amounts of fresh water would be at hand to use in the manufacturing process, and so that "used" water could be put back into the river. The law says the "used" water cannot be put back into the river unless it is neutralized. The pH level must fall within the range of natural waters so that living things are not harmed. You are responsible for disposing of the "used" water without causing an environmental problem.

CHALLENGE

Your challenge as a safety officer is to neutralize the pH level of the "used" water so that it matches the pH level of the river water. You and your partner will be given small amounts of the "used" water and the river water. Using these small amounts of water, you must determine how much base will be needed to neutralize 100 liters of the "used" water.

Discharges into the environment are regulated by law.

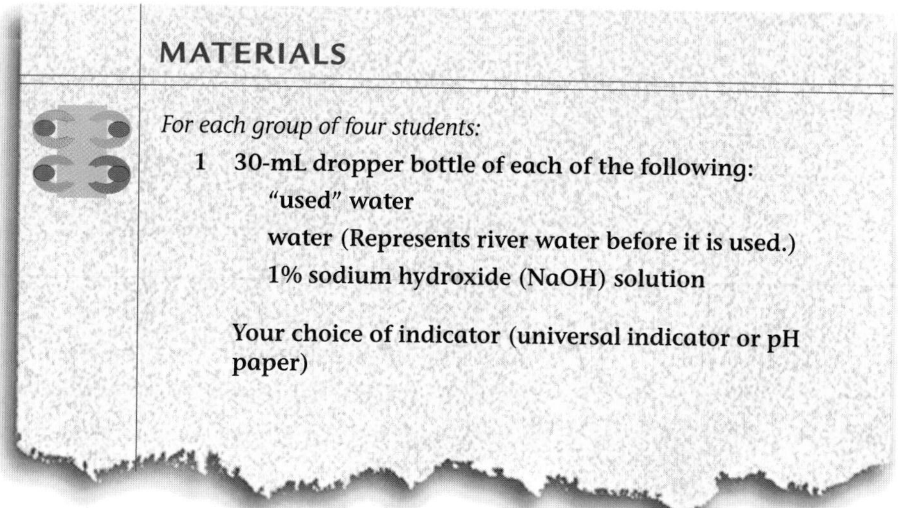

MATERIALS

For each group of four students:

1 30-mL dropper bottle of each of the following:
- "used" water
- water (Represents river water before it is used.)
- 1% sodium hydroxide (NaOH) solution

Your choice of indicator (universal indicator or pH paper)

SAFETY NOTE: **Do not taste or touch the solutions. Follow your school policy regarding the use of safety eyewear. Wash your hands after you complete the activity.**

PROCEDURE

Part One: Planning Your Investigation

- Keep careful notes of your procedure and your results. You will use them to write up a full report of your investigation.
- Set up a page for your investigation report in your science notebook.
- Think about how you might accomplish your task. Make notes in your investigation report.
- Share ideas with your partner. Working together, come up with a plan for finding out how much base is needed to neutralize a small sample of the acid waste water. Be sure to decide how many drops of "used" water you will test. Record your plan. Be sure to list all the materials you will need.

Part Two: Carrying Out Your Investigation

- Perform your investigation. If you make any changes to your original plan, be sure to record them.
- Remember to record all results.

ANALYSIS

- Based on your results, calculate how much base you would need to neutralize 100 liters of used water. Record your calculations.

Preparing a Written Report

As Safety Officer, you must write a report on your investigation. Start your report on a clean sheet of paper. Include your name and your partner's name, the date, and a title for your report. Your report should include the following:

- A statement of the problem you were trying to solve. In this case, include an explanation of why it is important to be sure to add the correct amount of base to the acid waste water.
- A description of the materials and procedure you used to solve the problem.
- A clear presentation of the results you obtained.
- An analysis of the results and your recommendation for neutralizing the waste. Be sure to include any calculations you performed. You should discuss any problems you had with the experiment, or any sources of error you may have had. Be sure to discuss water quality problems that could occur if you made an error in your experiment or your calculations.

13 Is Neutralization the Solution to Pollution?

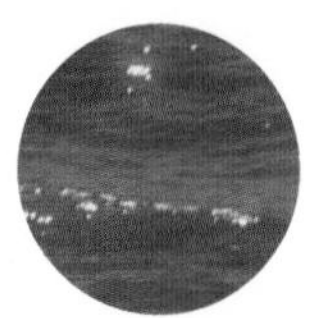

In Activity 12 you discovered that a base can be used to neutralize acidic wastewater. A pH test shows that the neutralized acidic wastewater behaves like distilled water. Does this mean the neutralized waste contains only water? It is important to know if neutralizing the water removed all the pollution.

CHALLENGE

Your challenge is to investigate the neutralized solution in order to find more evidence about the result of the interaction between the acid and base solutions.

How is wastewater treated in your community?

MATERIALS

For each group of four students:

2 SEPUP trays
2 stir sticks
2 30-mL graduated cups
2 paper clips
1 small bulb with attached battery harness
1 9-V battery
water
1 sugar packet
1 salt packet
1 30-mL dropper bottle of 1% hydrochloric acid (HCl) solution
1 30-mL dropper bottle of 1% sodium hydroxide (NaOH) solution
pH paper
2 droppers
1 8-ounce plastic cup half filled with water

SAFETY NOTE: **Do not taste or touch the solutions. Follow your school policy regarding the use of safety eyewear. Wash your hands after you complete the activity.**

PROCEDURE

1. Set up your investigation report. The table on page A-78 will help you record information.

2. Attach the battery harness and bulb to the 9-volt battery. Touch the two clips at the ends of each lead together briefly; notice how brightly the bulb shines.

   Note: To avoid wearing out the battery, do not allow the clips to continue to touch each other for long periods of time.

3. Set up your SEPUP trays according to the following table. Each pair of students should be responsible for one tray setup and testing. Each student will record all results.

Tray 1		Tray 2	
Cup A	10 mL hydrochloric acid	Cup A	10 mL water
Cup B	10 mL sodium hydroxide	Cup B	10 mL sugar-water*
Cup C	5 mL hydrochloric acid	Cup C	10 mL salt-water**

*Add a small packet of sugar to 10 mL of water and stir. Use all the sugar.

**Add a small packet of salt to 10 mL of water and stir. Use all the salt.

4. Use the pH paper and dropper method to find the pH of each solution. Make sure the solution in Cup C of Tray 1, is neutral. If it is not neutral, add the appropriate number of drops of hydrochloric acid or sodium hydroxide until it is neutral. Record the pH of each solution.

5. Attach a paper clip to each lead. These paper clips allow you to test the solutions without touching them directly to the harness clips. Test each solution for conductivity (ability to conduct electricity) by placing the two paper clips in the solution and observing the light bulb. Record all results.

   Note: Do not allow the paper clips to touch each other when you are testing a solution. Be sure to rinse the paper clips with water in the 8-ounce plastic cup between each test.

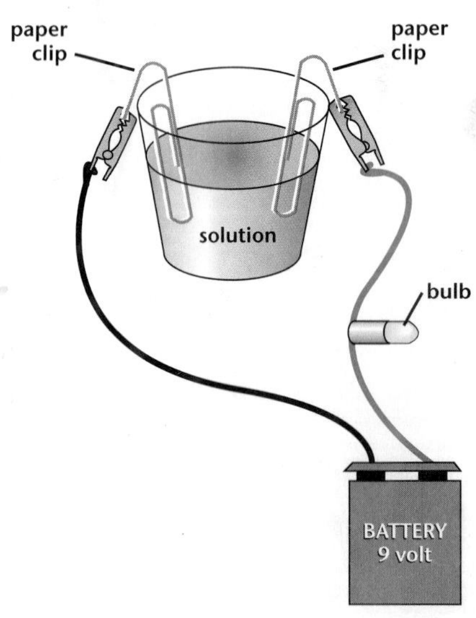

pH and Conductivity of Liquids

Solution	pH	Conductive?
Acid		
Base		
Neutralized acid + base		
Water		
Sugar-water		
Salt-water		

ANALYSIS

1. Which substances were neutral?
2. Which substances were conductive?
3. Did the neutral acid and base mixture behave like water? Explain.
4. Based on your results, what do you think the product of neutralizing an acid with a base could be? Explain your evidence for your idea.
5. How could you test your idea?
6. Carefully observe the demonstration your teacher will do. Use the information you obtain from the demonstration to support or revise your hypothesis about the neutral solution.
7. Based on everything you have learned about acid-base neutralization, do you think neutralization can be used as the only solution to acid-base pollution? Explain your answer thoroughly. Describe both the advantages and disadvantages and other factors that must be considered before reaching a conclusion.

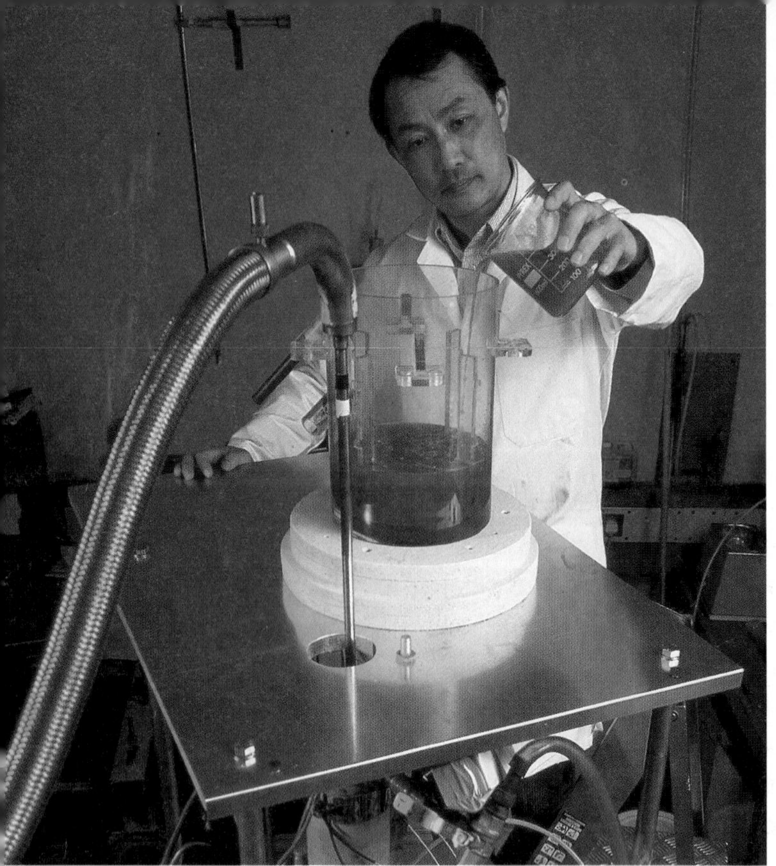

2 Investigating Groundwater

Over 75% of the people in the United States obtain their drinking water supply from water that travels beneath the Earth's surface. If you could dig a hole below your school grounds, you might hit water at 30, 60, or even 200 feet, depending on where you live. Long ago, this water fell as rain and soaked into the soil. Probably starting near mountains, or upstream of large rivers, the water has slowly traveled underground through layers of soil and rocks that act like sponges to soak it up. This water is called groundwater.

Silver Oaks is a town of 25,000 people not far from a large American city. It has a problem. Recent tests show that one well-water sample contains dangerous amounts of mercury—a heavy metal that can be stored in the human body. Mercury can produce many serious, long-term health effects. In this section, you will act as a detective to discover the source of the mercury contamination. After locating the source, you will propose some methods to clean it up. With your help, Silver Oaks can solve its water contamination problem.

14 Trouble in Silver Oaks

The Silver Oaks Story

Most of the time we take a safe drinking water supply for granted. What happens when the drinking water supply is threatened by contamination?

CHALLENGE

Think about how you would react if you lived in Silver Oaks and your drinking water supply were threatened by contamination.

Treatment of waste water often includes aeration and settling ponds.

THE SILVER OAKS STORY

Carla and her grandparents live in Silver Oaks. Carla likes living here because she can walk to the large community park and meet her friends. She learns about the town's history when she walks in the early evenings with her grandfather through the city cemetery to see the old tombstones near the lake. The town has changed dramatically since it was founded in the early 1800s. Areas that were once wooded are now used for factories, schools, businesses, residential subdivisions, and a landfill.

Willow Lake is in the northeastern section of Silver Oaks. The lake once provided the water necessary for many of the industries that operated in Silver Oaks. Now many of these plants have closed, leaving deserted factory buildings. The people who live in central Silver Oaks, and in the older homes in Golden Oaks, use water piped to their houses from the Silver Oaks Municipal Water District. The water district draws its water from wells and from Willow Lake. This water is treated to meet all federal standards before it is piped to the customers. The people who live in outlying areas still use water from their own wells.

Carla lives in Silver Oaks Estates, a section of Silver Oaks that is not yet hooked up to the Silver Oaks Municipal Water District. Recently, Carla and her family have noticed a funny smell near their well and suspect that it may be contaminated. Just to be safe, her family has started drinking bottled water. In addition, Carla's grandmother has noticed more outbreaks of "flu" in her neighborhood this year than in previous years.

Discuss with your group important pieces of information, or evidence, that you have learned so far about the possible problem in Silver Oaks.

Inside a water treatment plant

Carla thinks that the problems people are having might have something to do with the water. She is studying water testing in her science class at school. She discusses her concerns with her grandparents, and they decide to collect water samples from her well, Willow Lake, and Fenton River. The family arranges to have the samples tested by a local water testing company. The tests of the three samples of water find that there is a concentration of 3 ppb (parts per billion) mercury in the well water. The maximum recommended level is 2 ppb. Carla remembers from study about cholera in her science class that a map might be helpful. As she begins to record the information on the map shown on page A-85, she begins to think she could help solve the problem.

After completing the map, Carla and her grandparents talk with the health commissioners of Silver Oaks about the test findings. The health officials are impressed with the information that Carla collected and the map that she made. They are concerned about the source of contamination and how quickly it may be spreading through the underground water supply. The health officials ask Carla and her classmates to help decide where wells should be drilled for testing.

QUESTION

Imagine that you live next to Carla and her grandparents. What would you do about the problem? Be sure to describe your concerns and any steps you would take after learning about the problem with the Silver Oaks water.

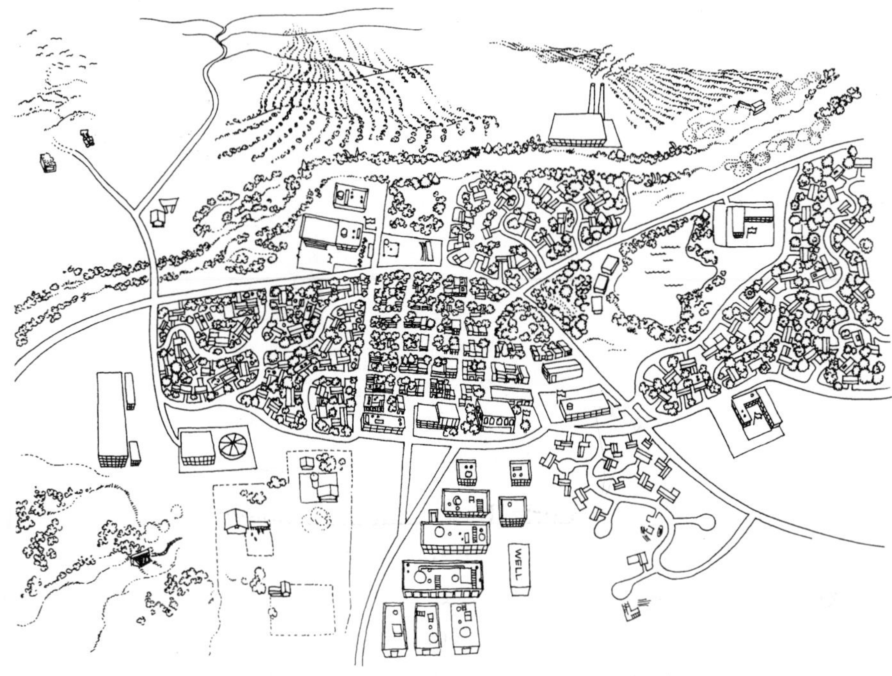

Line drawing of Silver Oaks showing elevation and contour

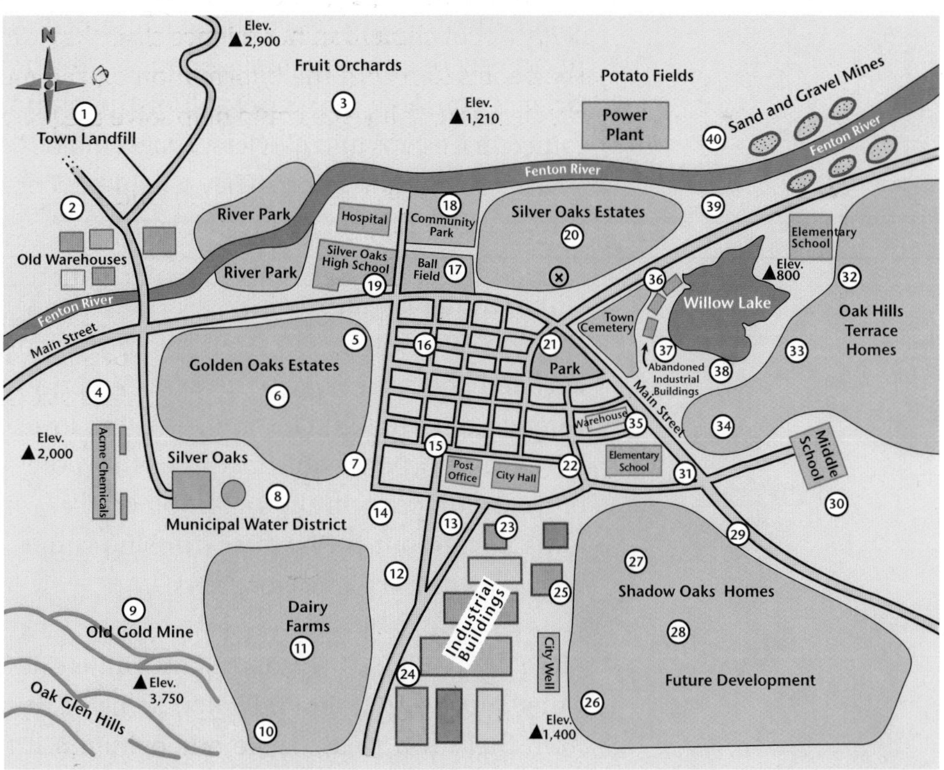

Map of Silver Oaks: the "X" marks Carla's well

Earth's Groundwater Supply

You may think that most of our drinking water comes from sky blue lakes, crystal clear rivers, and large reservoirs. But these sources are small compared to the abundant water supplies found underground. This water supply is called groundwater.

CHALLENGE

Use the reading and diagram to learn how groundwater can become contaminated.

WATER IN THE GROUND

Think of all of Earth's water, both fresh water and salt water. Imagine that all of it is contained in a 2-liter soft drink bottle. If you took out all the salt water, the remaining water would represent fresh water. The fresh water could be contained in a small glass about 52 mL in size, or approximately 2 ounces—the amount of water in a popsicle! If you now took away all of the groundwater and water contained in ice caps and glaciers, you would be left with all of the Earth's water in lakes and rivers. They would be represented in our model by 0.28 mL, or less than 5 drops of water!

Groundwater is water that travels down through the soil, like hot water through coffee grounds. It may fill up layers of soil, or it may be stored between spaces in layers of rocks. When water flows underground through a **permeable** layer of rock, the layer is known as an **aquifer** (AH-kwi-fer) (permeable means having pores or openings that liquids or gases can pass through). As you can see, groundwater makes up a large part of our fresh water drinking supply compared to those sky blue lakes and crystal clear rivers!

More than 75% of the people in the United States obtain their drinking water supply from groundwater. Some of these people get the water from their own wells, while others obtain it from municipal (city or town) water companies. The rest of the country's water comes from surface water sources, such as lakes, rivers, and streams.

By its very nature groundwater is never "pure" water. Water from various sources, such as rain, snowfall, and agricultural irrigation, carries dissolved and suspended materials below the surface. It was once thought that the soil acted as a natural filter and removed dangerous substances before they affected groundwater. Unfortunately, that is not entirely true. Today, we find that our groundwater can be contaminated by such natural causes as radioactive minerals (radium) and gases (radon), toxic substances given off by plants and animals, and the action of microbes like giardia and amoebas. The human population adds to this contamination problem by the use of pesticides and fertilizers, leaking underground storage tanks, poor sewage (septic tanks) and waste disposal practices, industrial discharges into lakes, rivers and streams, and air pollutants that dissolve in rain water.

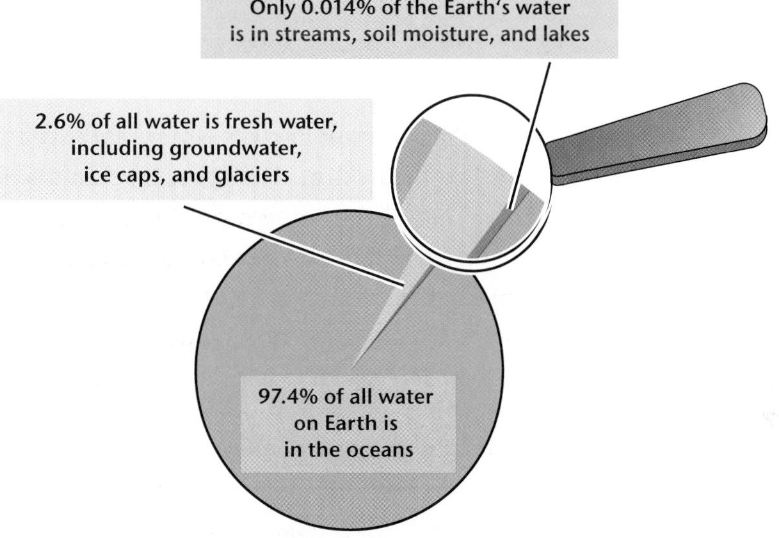

Groundwater and surface water (such as streams and rivers) are interconnected. Surface water helps to replace groundwater that has been removed for human use, while groundwater helps keep streams and rivers flowing when there is a drought. Groundwater makes possible a year-long water supply, ensuring the health of lakes, rivers, streams, and oceans.

QUESTIONS

1. What is groundwater? Why is it important?
2. Does the water you drink come entirely, in part, or not at all from groundwater?
3. How can surface water become contaminated? How can ground-water become contaminated?

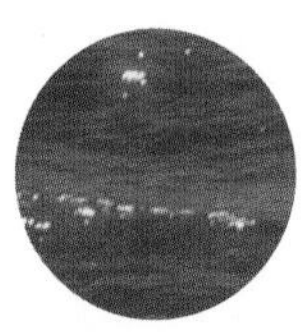

Reviewing the Evidence

Silver Oaks has a water contamination problem. Sometimes it is helpful to organize information about a problem before continuing an investigation.

CHALLENGE

Make a table in your science notebook to summarize the evidence and information about this mystery. Be sure to look at the map and drawing of Silver Oaks on page A-85.

A water treatment plant

PROCEDURE

1. Prepare a table in your science notebook like the sample below. Use a full page. Use what you have read about Silver Oaks to list evidence in the first column about the water contamination problem.

2. Use the second column to list any ideas you have that might help explain the mystery.

3. Use the third column to list questions you need to ask about water, mercury, Silver Oaks, or other information to help solve the contamination mystery.

Silver Oaks: Review of the evidence

Evidence What facts do I know?	Ideas Ideas I have about the location	Questions What do I still need to know?

15 Water Movement Through Earth Materials

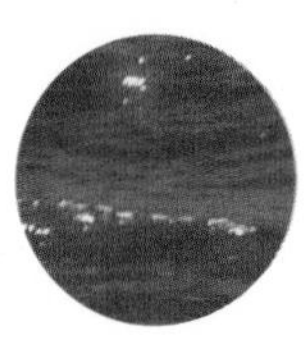

Liquids and Solids

By investigating water movement through solid materials such as clay, sand, and gravel, you will begin to understand how mercury contamination may have traveled to the contaminated well in Silver Oaks.

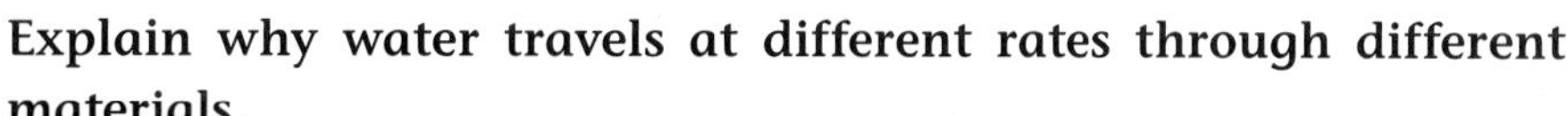

CHALLENGE

Explain why water travels at different rates through different materials.

INTRODUCTORY QUESTIONS

1. List three examples of a liquid moving through a solid other than soil. Think of examples that occur in your daily life.
2. What determines how fast a liquid moves through a solid? Give two explanations.
3. Make a drawing that illustrates how a liquid travels through a solid.

PROCEDURE

1. Your teacher will demonstrate how water moves through some earth materials. Record the names of the materials in Tubes A, B, and C on a copy of the data table that appears on the opposite page.
2. Rank the materials from 1 to 3 in the order you think they will transmit water. Use a "1" for the fastest and a "3" for the slowest. Record your predicted ranking on the data table.
3. Predict how long you think it will take for the water to reach the bottom of each tube. Record your predicted time on the data table.
4. Observe the demonstration of the water poured into each tube. Record on your data table the actual time it took for the water to reach the bottom.
5. Using the actual time results, complete the data table for the actual rank order of water flow through the earth materials.

Water Movement Through Earth Materials

Tube	Material	Predicted Rank	Predicted Time	Actual Rank	Actual Time
A					
B					
C					

ANALYSIS

1. Explain why the water moved faster through some types of earth materials than through others.

16 The "Ins and Outs" of Groundwater

The Global Water Cycle

The world's freshwater supply is constantly recycled through the global water cycle.

Trace the path of fresh water as it moves through the water cycle.

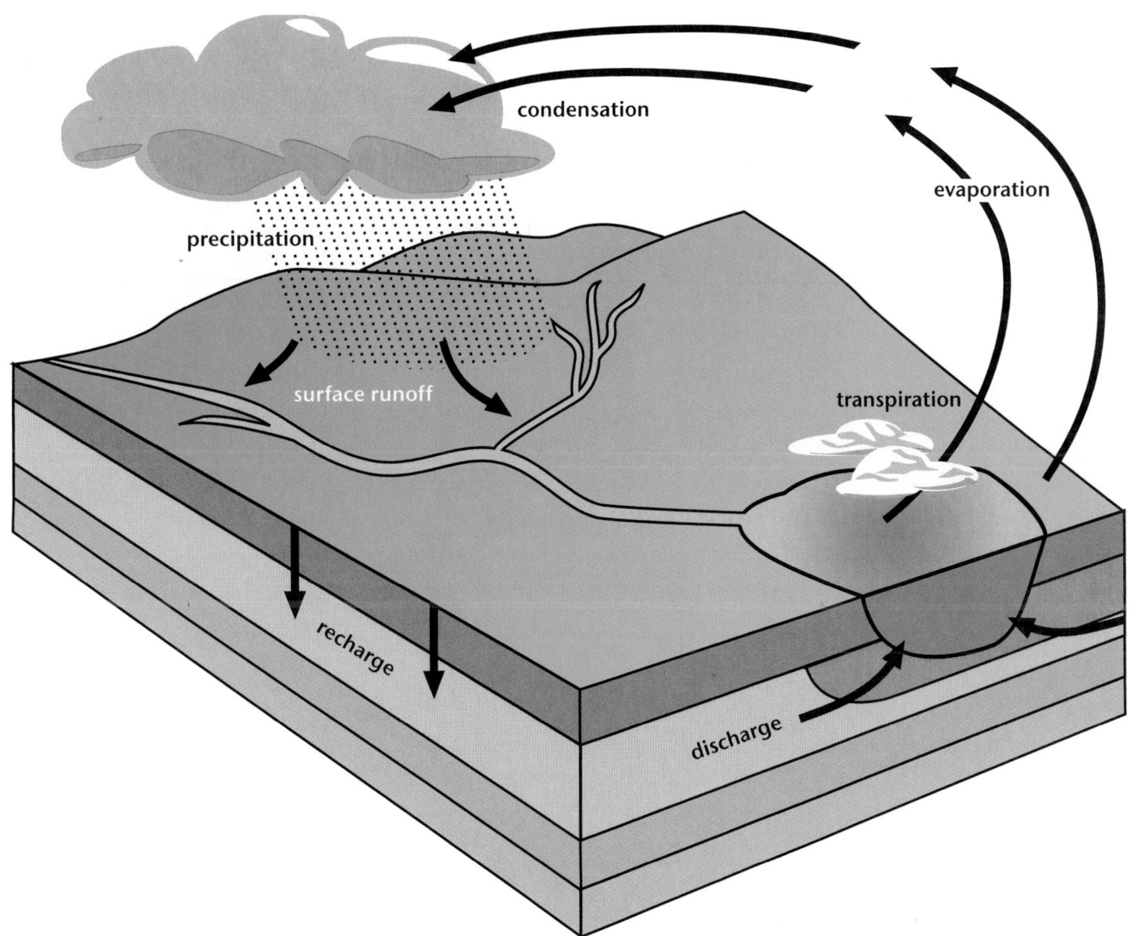

THE WATER CYCLE

All the fresh water in the world's lakes and creeks, streams and rivers, represents less than 0.01 percent (1 part in 10,000) of the Earth's total store of water. Fortunately, this freshwater supply is renewed when water vapor from the atmosphere returns to Earth as rain or snow (precipitation).

The global water cycle has three major pathways: precipitation, evaporation, and vapor transport. Water precipitates from the sky as rain or snow. Most of this water falls into the oceans, but some of it falls on land to renew the freshwater supply. Some fresh water flows from the land to the sea as runoff or groundwater. The water that evaporates from the sea and from the surface water on land enters the atmosphere as water vapor. The water vapor is then transported, or carried, over the land by atmospheric currents (wind). The cycle begins again as the water condenses and precipitates from the sky to the land in the form of rain and snow.

When surface water sinks into the ground, it renews or recharges the groundwater supply. This process can also go in the opposite direction. Groundwater comes to the surface at natural springs.

As water moves through the water cycle, particles and materials dissolved in the water can be carried along. When the water evaporates, these things are left behind, and the water enters the atmosphere as pure water vapor. Once the water vapor is in the atmosphere, however, it can be contaminated by gases and particles that result from both human activity and natural events such as volcanoes. As water runs over the land, it picks up additional materials that are eventually carried to rivers, lakes, and the ocean.

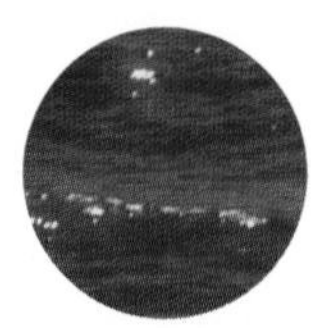

Understanding Groundwater

The ground below your feet can act as a giant sponge. It can absorb the rain and transport it deep underground. There, the water may encounter layers of rocks that can absorb it or hold it back from moving to deeper layers. Eventually all water traveling underground returns to the ocean. How fast does this process happen?

CHALLENGE

Does water travel at the same speed through different types of soil and rocks? In this activity you will explore the answer to this question and discover how earth materials can permit or restrict the flow of water.

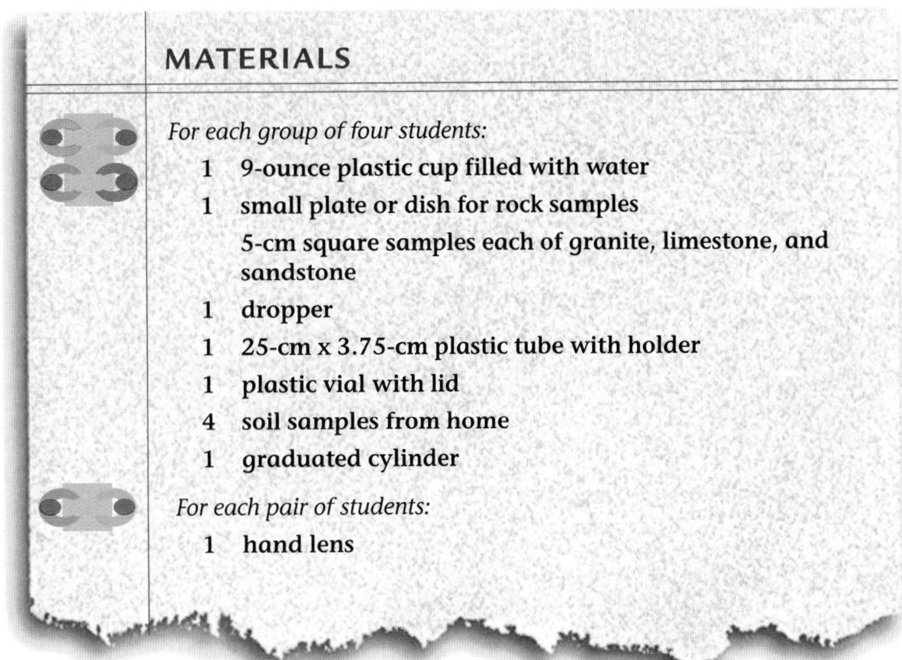

MATERIALS

For each group of four students:

- 1 9-ounce plastic cup filled with water
- 1 small plate or dish for rock samples
- 5-cm square samples each of granite, limestone, and sandstone
- 1 dropper
- 1 25-cm x 3.75-cm plastic tube with holder
- 1 plastic vial with lid
- 4 soil samples from home
- 1 graduated cylinder

For each pair of students:

- 1 hand lens

PART ONE: SEPARATING SOIL PARTICLES

Procedure

1. Fill a plastic vial half full with a "Standard Soil Sample."
2. Add water to the sample until the water level is near the top of the vial.
3. Cap and shake the vial.
4. Allow the contents to settle out.

Analysis

1. Draw a picture of the layers formed in the sample. Label them.
2. Why do you think the layers are in this order? Explain.

PART TWO: TESTING SOIL SAMPLES

SAFETY NOTES: **Do not chip or break the rocks. Some rocks may have sharp edges.**

Procedure

1. Examine your soil sample with a hand lens. Describe what you see in your data table. Use a table like the one on the next page.
2. Fill the large plastic tube with your soil sample, and tamp it down lightly so that the soil settles.
3. Predict how long it will take for water to travel to the bottom of the container. Base your prediction on the observations you made in the last activity comparing samples of gravel, sand, and clay, as well as your observations of what your sample contains. Record your predicted time.
4. Pour 30 mL of water into the top of the tube and record the amount of time it takes for water to first reach the bottom of the tube. Record your actual time.

Observations of Soil Samples

Sample	Description of Sample	Predicted Time	Actual Time
1			
2			
3			
4			

Analysis

1. Is your soil sample more like the sand, gravel, or clay that you observed in the last activity? Or is it a combination of all three?
2. Observe the results of other groups and write a short paragraph in your science notebook comparing their samples with yours.

PART THREE: TESTING ROCK SAMPLES FOR PERMEABILITY

Procedure

1. Identify the three rock samples at your table. Use the hand lens and record your observations in a table like the one shown below.
2. Add one drop of water to the surface of each rock sample and record your results in the data table.
3. Use a scale from 1 to 3 to rate how quickly water flows into each rock. Let "1" represent very slow water flow, and "3" very fast water flow into the rock.

Observation and Testing of Common Rocks

Rock	Origin	Description	Permeability (Use scale 1–3)
Granite	deep underground from cooling magma		
Sandstone	from layers of sand deposited in water		
Limestone	from layers of shells deposited in water		

Analysis

1. Which rocks were most permeable to water? Which rocks were least permeable? What is your evidence?

PART FOUR: GROUNDWATER MOVEMENT

Surface water trickles down through the earth at a rate of several inches to several feet per day. This water may reach layers of rock such as granite, sandstone, or limestone. In permeable rocks, the water collects like a sponge full of water at various depths below the surface, to form an **aquifer**. This is groundwater. The diagram below shows the movement of groundwater and some of its sources. The arrows represent the direction of water movement. Some earth layers do not permit the flow of water. These are called **aquitards**.

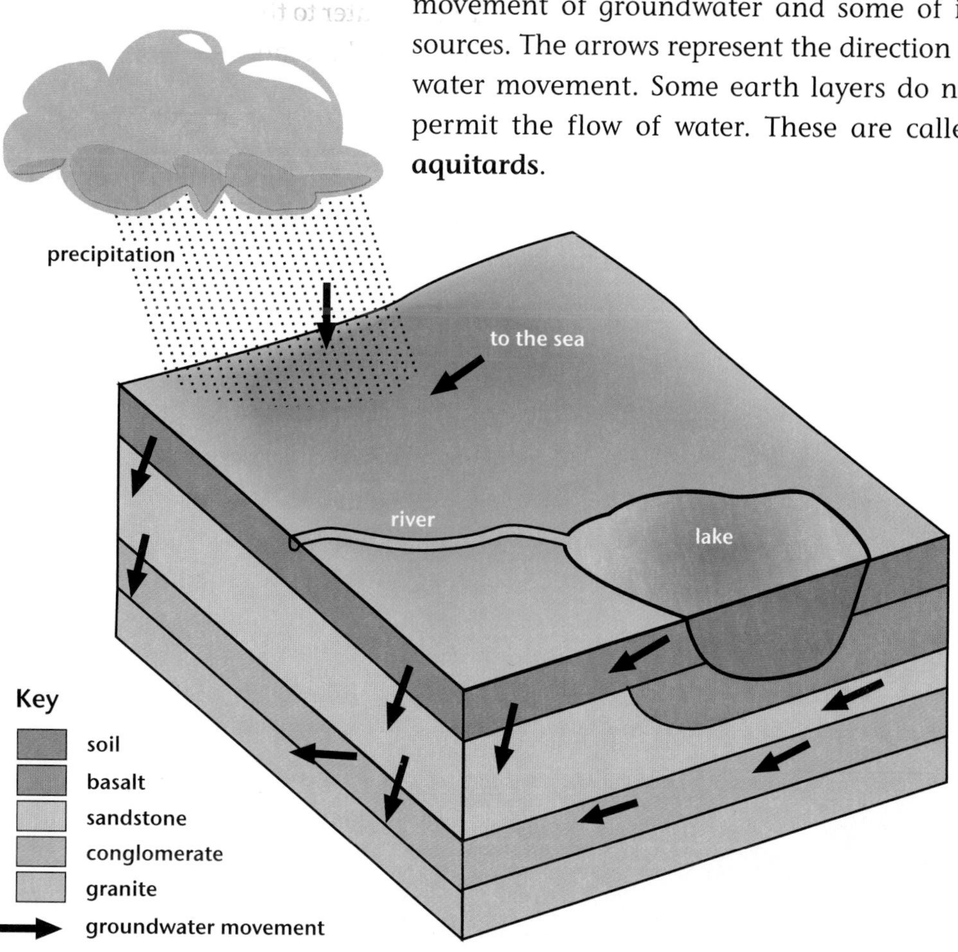

QUESTIONS

1. What is an aquifer? List the aquifers on the diagram by using the name of the layer of material they are made of.
2. What is an aquitard? List the aquitards on the diagram by using the name of the layer of material they are made of.
3. Describe the flow of the water from the surface of the Earth through the aquifers to the conglomerate level.

All About Mercury

Carla and her family fear that mercury may be contaminating their drinking water. This reading will help you understand some of the interesting properties of mercury, and how it can be highly toxic to humans.

Use this reading to learn more about how mercury may have contaminated the Silver Oaks groundwater and about the health risks of mercury.

MERCURY AND ITS COMPOUNDS

Mercury, which is also commonly called quicksilver, is one of only three liquid elements. It is the only metal known to take a liquid form at room temperature. Along with gold, silver, and copper, it is one of the few elements that is found in nature in its pure form. (An **element** is one of a few fundamental substances—there are just over 100—that contain only one kind of atom. All of the matter in the universe is made up of these elements, alone or in combination with other elements.)

Ancient people believed mercury had magical healing powers and used it to cure diseases. In fact, its toxic properties make it an effective, if dangerous, disinfectant. Until recently, mercury was mixed into a compound called mercurochrome that was used to kill bacteria in cuts. Seeds were coated with mercury compounds to prevent them from molding when they were placed in damp soil. Mercury was also used to process the felt that went into making hats. People eventually learned of its highly toxic properties. The expression "mad as a hatter" comes from the observation that people who used mercury compounds to make hats showed symptoms of mental illness. This was caused by the mercury they worked with.

Microscopic bacteria transform liquid mercury into a toxic form called methylmercury.

Microscopic plants and animals pick up methylmercury as they feed on bacteria.

Small fish eat large numbers of these plants and animals.

These fish may then be eaten by larger fish.

In this way, mercury is concentrated, or magnified, as it travels through the food chain.

Note: Drawing is not to scale.

Mercury has many more uses today. It has long been used in thermometers because it expands uniformly when heated. It is also used in barometers to measure air pressure. Mercury is used in the manufacture of plastics and pesticides. In your home, mercury is found in the fluorescent tubes used for lighting and in some thermostats, where it acts as a conductor of electricity for a switch that senses temperature changes.

Mercury is a solvent for some metals, that is, they dissolve in it. As a result, it is used to remove tiny bits of gold that would otherwise be too hard to separate from the gold ore (rock) in which they are embedded. When mercury is mixed with gold-containing ore, the gold dissolves in the mercury, forming what is called an amalgam (ah-MAL-gam). Then the mercury is boiled away, leaving the gold behind. Dentists use a mixture of silver and tin dissolved in mercury for fillings.

Mercury can produce both acute and chronic toxicity. This means it can make people very sick immediately (acute toxicity) as well as causing long-term damage to health (chronic toxicity). Mines, landfills, and incinerator gases put mercury and mercury compounds into the environment. When people inhale very low doses of mercury vapors, the metal can be stored in the body, resulting in chronic mercury poisoning. The compound mercury chloride is quickly absorbed into the body. Very small amounts of this compound can result in an acute toxic dose and death.

When mercury is released into the atmosphere, it may make its way to lakes and rivers. There, bacteria transform it into a toxic form called methylmercury. Small microscopic plants and animals pick up this methylmercury as they feed. Small fish eat large numbers of these plants and animals. If there is mercury in the water, the concentration may increase in these fish from parts per trillion to parts per billion. These fish may then be eaten by larger fish, such as pike, tuna, and swordfish. In these large fish the mercury may now be hundreds of parts per billion. In this way, the mercury is concentrated, or magnified, as it travels through the aquatic food chain. People who regularly eat fish that have high concentrations of mercury, your chances of mercury poisoning are greatly increased.

Most people do not eat enough fish for the mercury levels in their bodies to rise to dangerous levels. Therefore, the average person is at no great risk from exposure to mercury. But in countries where fish intake is high and there is significant mercury contamination, or in areas of high local mercury pollution, people have experienced mercury poisoning.

The source of nearly 50% of the methylmercury in the environment is combustion (burning). Industrial plants and waste incinerators that process household wastes containing mercury (batteries, electrical switches, and fluorescent lights) emit a small, but significant, amount of mercury gas into the atmosphere. This gas dissolves in rain and is transported to bodies of water. Over half of the mercury contamination in a Canadian lake was traced to the atmosphere, and the other half came from natural mineral sources. Acid rain, which is caused by industrial air pollution, carries the mercury vapor into lakes and streams and then provides an acid environment that is friendly to the bacteria that form methylmercury.

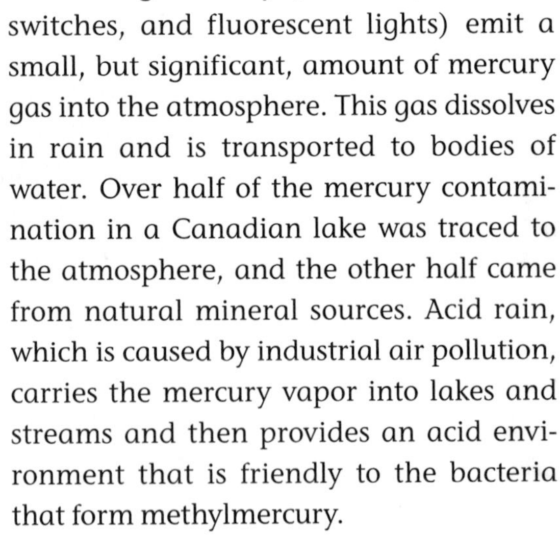

Mercury often enters the air during combustion.

QUESTIONS

1. Now that you have read about mercury, what concerns do you have about the contaminated water that was found in the Silver Oaks well?

2. Based on the reading, where do you think the mercury in the well water in Silver Oaks might have come from?

Modeling Contamination Plumes

By traveling in groundwater, toxic substances can contaminate an area when they leak or are spilled from a very concentrated source of material. This happens when a barrel or an underground storage tank leaks. Spills and leaks such as these are called **point sources** of contamination. When a toxic material is spread over a large area, as when pesticide is sprayed from an airplane, the source is called a **non-point source** of contamination. An underground area where contaminants have been spread by groundwater is called a **contamination plume**.

CHALLENGE

Your challenge is to determine how point and non-point sources of contamination change the shape of an underground plume. You will investigate two models of plumes to find out. The sand in this activity represents an aquifer, and the food coloring represents a contaminant, such as mercury.

Area or point source?

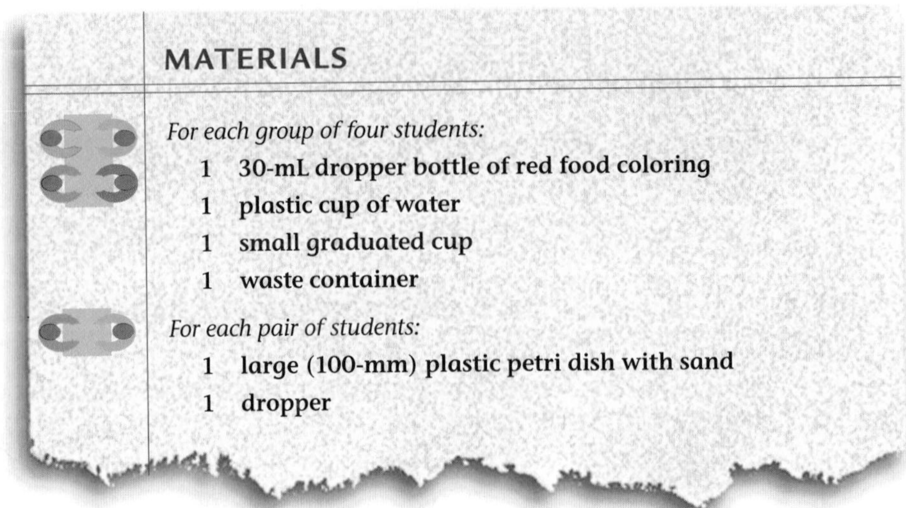

MATERIALS

For each group of four students:

1 30-mL dropper bottle of red food coloring
1 plastic cup of water
1 small graduated cup
1 waste container

For each pair of students:

1 large (100-mm) plastic petri dish with sand
1 dropper

PROCEDURE

1. Each group of four students will divide into two pairs. Each pair will need a petri dish with sand. In your science notebook carefully note observations and include drawings of what you see happening. You are also responsible for recording the other pair's observations in your science notebook.
2. One pair will do Part One, and the other pair will do Part Two of the investigation.

SAFETY NOTE: **Food coloring can stain clothes and work surfaces. Immediately clean up any spills with water.**

Part One: Point Source Contamination

One pair in your group does this part.

1. Use the dropper to slowly add water to the sand in the petri dish until all of the sand is damp, from top to bottom. If you add too much water, carefully pour it off into the dishpan provided by your teacher.
2. Lean one edge of your petri dish against a closed book so that the dish is slanted. It should be supported by the book and not fall away from it.

3. Make a small hole in the sand in the middle of the dish. Add two drops of food coloring to the hole and cover it gently with sand. The drops of food coloring represent mercury.
4. Slowly add 20 drops of water to the sand at the top end of the petri dish and carefully observe what happens for the next 2 minutes.
5. Thoroughly examine the bottom of the dish and record all of your observations in your science notebook.
6. Clean up as directed. Do not discard sand in the sink. Use the container provided!

SAFETY NOTE: **Food coloring can stain clothes and work surfaces. Immediately clean up any spills with water.**

Part Two: Non-point Source Contamination

The other pair in the group does this part.

1. Use the dropper to slowly add water to the sand in the petri dish until all of the sand is just damp, from top to bottom. If you add too much water, carefully pour it off into the dishpan provided by your teacher.
2. Lean one edge of your petri dish against a closed book so that the dish is slanted. The dish should be supported by the book and not fall away from it.
3. Add a drop of food coloring to 15 drops of water in the graduated cup. Mix. This solution represents a more diluted form of mercury that will spread over a wide area.
4. Place four small holes in the sand across the middle of the dish. Place two drops of the diluted food coloring in each hole. Gently cover the holes with sand.

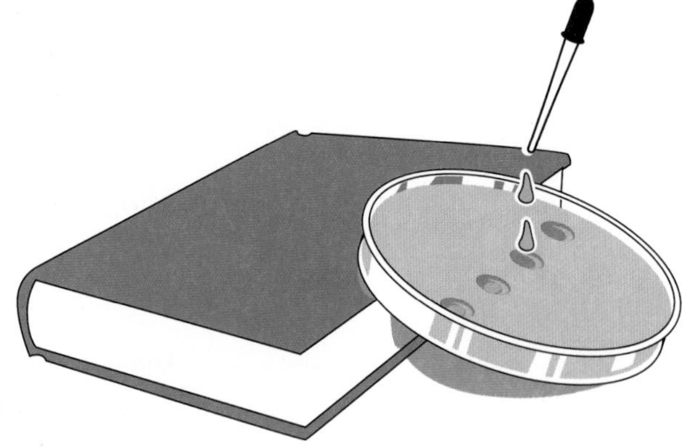

5. Slowly add 30 drops of water in a line back and forth across the top of the petri dish.
6. Wait two minutes, then carefully examine the bottom of the dish and record all of your observations in your science notebook.
7. Clean up as directed. Do not discard sand in the sink. Use the container provided!

ANALYSIS

Summarize the similarities and differences in how contamination from point sources and non-point sources moves through aquifers.

FACTS ABOUT GROUNDWATER

- Approximately 75% of the United States population depends, at least partially, on groundwater for drinking water.
- An estimated 95% of rural households get all of their drinking water from groundwater.
- The federal government estimates that 2% of the nation's groundwater is polluted from point sources. This is actually more of a threat than the low figure might indicate because the pollution tends to be near high population areas and thus affects more people. Also, this figure does not include non-point sources of pollution such as agricultural and urban runoff.
- According to the United States Environmental Protection Agency (1998), about 10 percent of all ground water public water supply systems violate drinking water standards for biological contamination from time to time. Approximately 74 pesticides, a number of which are carcinogens, have been detected in the ground water of 38 states.
- Infections from **pathogenic** (disease-causing) bacteria, viruses, and parasites cause about 50 times as many cases of acute illness as chemical contamination of water.
- Apart from microbes, virtually every contaminant causes an **acute response** (makes people sick immediately) only if the substance is present in large amounts. In the small quantities in which most contaminants are present, the main risks are of chronic, low-grade illnesses which are hard to diagnose, and of cancer, which develops over years and is often hard to connect directly to any one cause.

Developing Your Well Testing Plan

In this and the next activity you will help the people of Silver Oaks. The community wants to determine the source of the mercury contamination and to measure how far it has spread. The first step is to drill three test wells and to test water drawn from them to find out how much mercury is in the groundwater.

CHALLENGE

Your challenge is to decide on the three locations where the water should be tested first.

PROCEDURE

1. You know that the well marked with an "X" on the map is contaminated with mercury levels that are higher than the federal safety standards. The health commissioners of Silver Oaks have identified 40 possible sites for drilling test wells. However, the community only has enough money to drill 12 test wells. Using clues from the Silver Oaks story, information from the Silver Oaks map, and your notes on groundwater movement, decide which three wells you think would be most important to test first.
2. Record the three wells you have chosen and your reasons for choosing them in your science notebook. Be prepared to explain your choices.

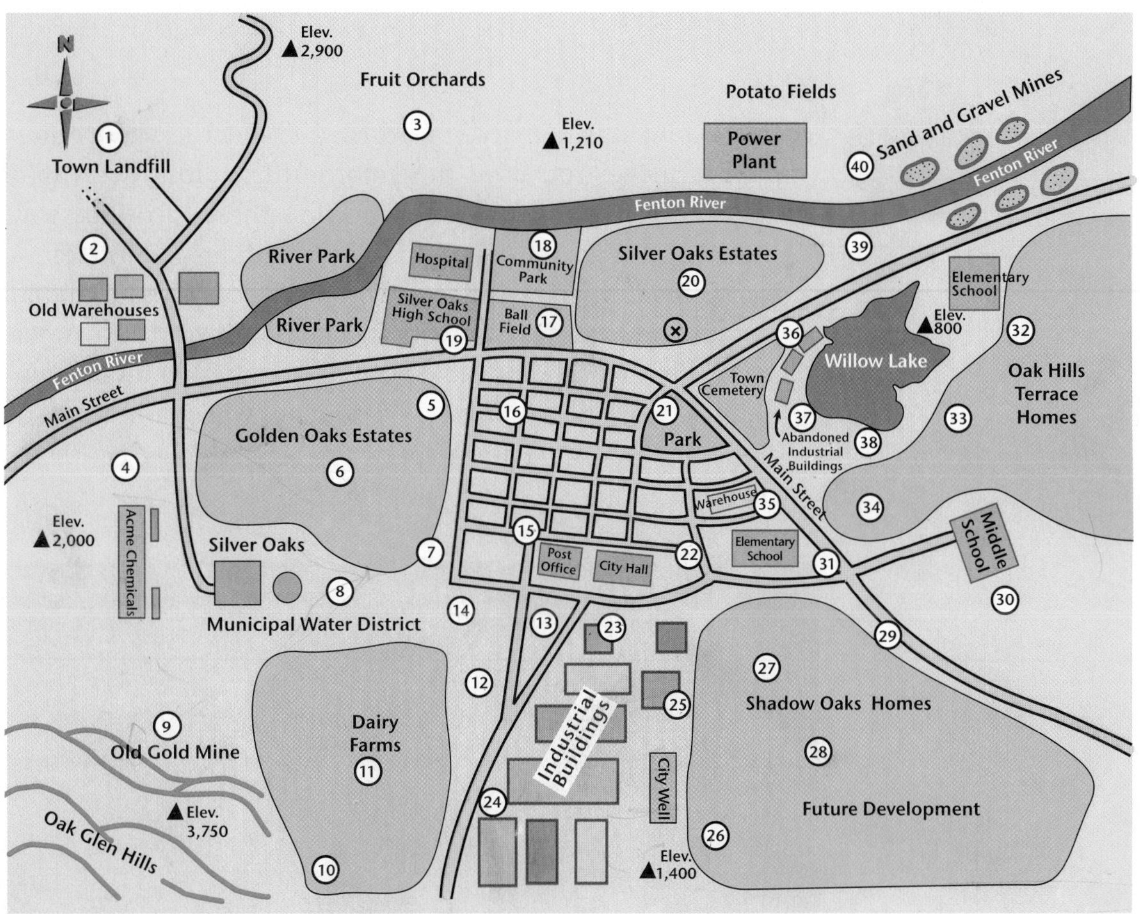

Map of Silver Oaks. The "X" marks Carla's well.

18 Testing the Waters

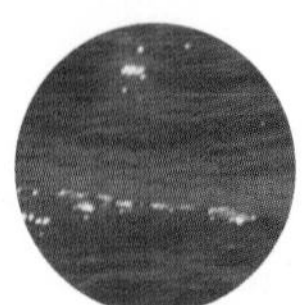

In the previous activity you learned about the way contamination plumes move. You used this information, along with information about Silver Oaks, to choose the first three locations where wells should be drilled and tested. In this simulation, you will test well-water samples to determine the source of the mercury contamination and to measure how far it has spread. Remember, Silver Oaks' funds are limited. Only 12 of the possible 40 well sites can be tested. You will test three wells at a time, and use the evidence you obtain from each phase of testing to plan the next phase.

Drilling test wells.

GROUP CHALLENGE

Working as a group, your challenge is to test 12 wells, three wells at a time, to determine the source of the mercury contamination and to find out how far it has spread through the aquifer below Silver Oaks. You will use this information to develop an "underground" map of the contamination plume.

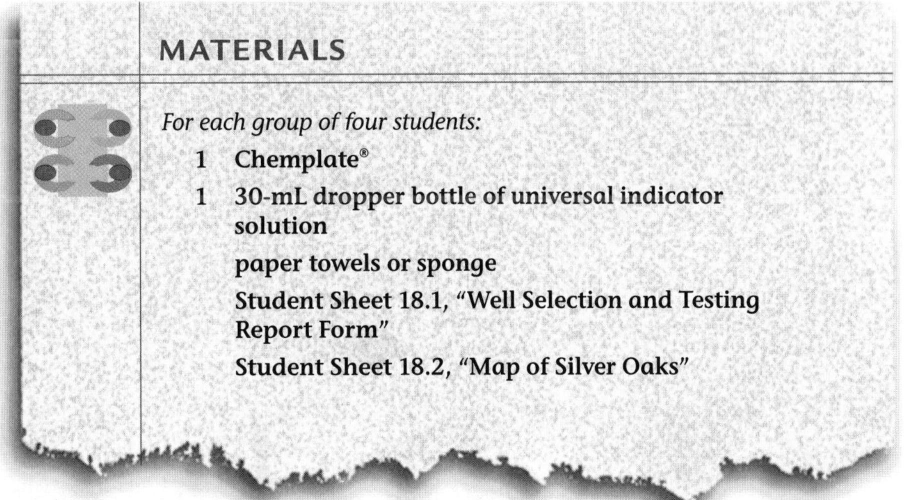

MATERIALS

For each group of four students:

1 Chemplate®

1 30-mL dropper bottle of universal indicator solution

paper towels or sponge

Student Sheet 18.1, "Well Selection and Testing Report Form"

Student Sheet 18.2, "Map of Silver Oaks"

SAFETY NOTE: **Wear safety eyewear and do not let the chemicals touch your skin. The test used is a simulation. There is no mercury in the samples.**

PROCEDURE

1. Take out your science notebook and turn to the page where you recorded your ideas about which three wells should be tested first. As a group, discuss the choices each of you has made. When you agree about which three wells to test first, record the three wells you chose and your reasons for choosing them in the Well Selection Table on the form on Student Sheet 18.1. Be sure to explain your reasons for your choices thoroughly.
2. Test the three wells for mercury. Each well is tested by adding 5 drops of the well water to one drop of universal indicator in the Chemplate.
3. Record the results of your test in the Well Testing Results Table. The chart at the bottom of this page tells you how to use the colors obtained with the universal indicator to determine the concentration range of the mercury and the code of the well.
4. You will continue to test the wells in groups of three until you have tested 12 wells and completed the report form.

Concentration Range Chart for Well Samples

Color	Concentration Range	Code
yellow-orange	not detected—less than 0.1 ppb	1
yellow	0.11 ppb–0.8 ppb	2
green	0.81 ppb–4 ppb	3
blue-green	4.1 ppb–32 ppb	4
blue	more than 32 ppb	5

19 Mapping It Out

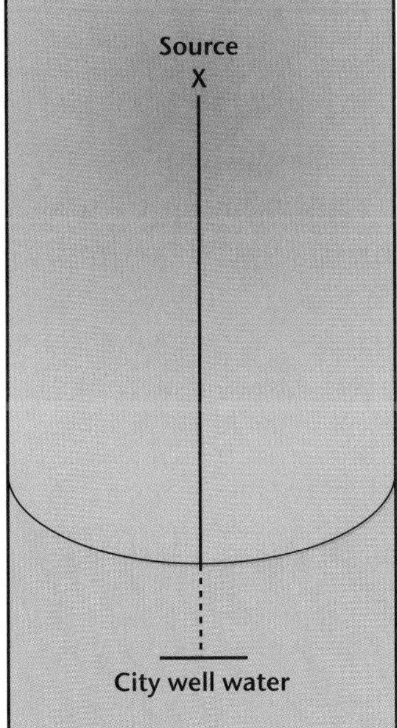

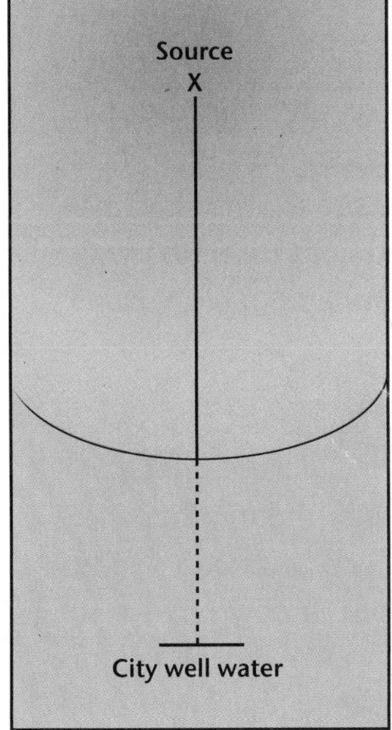

Scale
1 cm = 100m
0 m 200 m

The two maps at left show the contamination distributions of the same area drawn by two different people. The two tested different wells. They have marked the distance of the safe boundary from the source of the contamination. They have also marked the distance from the boundary to the city water wells. They know that the source area was contaminated only one time, and this contamination occurred five years ago.

CHALLENGE

Find out how quickly the contamination will reach the city water well.

QUESTIONS

1. Look at Map A. How far has the pollutant traveled in 5 years? How far has it traveled (on average) per year?
2. Look at Map B. How far has the pollutant traveled in 5 years? How far has it traveled (on average) per year?
3. If the pollutant continues to travel at the same average speed, how long will it take to reach the city wells?

   Map A: It has to travel ____ meters at a rate of ____ m/yr. It will take ____ years.

   Map B: It has to travel ____ meters at a rate of ____ m/yr. It will take ____ years.
4. Explain how two people's data could lead to such different predictions for when the contamination plume will reach the wells. Try to think of more than one possible explanation.
5. If you were a health officer for the city and these were the only plume maps available, which map would you use to make decisions about using water from the city wells? Explain.

Clean-Up Methods

There are several methods available for cleaning up contaminated groundwater.

Read about possible methods for cleaning up the groundwater contamination in Silver Oaks. Think about which method might work best. In Activity 20 you will have a town meeting to select a method.

METHODS PROPOSED TO CLEAN UP CONTAMINATION

Containment

Surround and cover the area with clay. Drill test wells and install pumps at selected points in the path of the plume. Monitor the mercury concentration in the groundwater so that the wells can be shut down if the levels are too high. This method is quick and relatively inexpensive. It minimizes air pollution, has been used for this purpose in other areas, and poses little danger to the public. Disadvantages include the need to monitor the water as long as the wells are used, and issues of worker safety while the clay cap is being installed. There are sometimes problems in choosing the correct site for the clay layer. In some cases, well pumps have actually pulled contaminated water from other areas into the aquifers as water is pumped out of them.

Removal

Use a pump to remove the contaminated water from the aquifer for treatment. This can be done quickly and has been used in other areas with relatively little danger to the public. However, pumping will not immediately remove contaminants trapped in sediments. Also, the treated residue may be classified as a hazardous waste and will require additional disposal procedures.

Excavation

Dig out the earth materials in the plume area and transport them to a hazardous waste site for disposal. This process has been used successfully in other areas and does remove material that is highly hazardous. However, it also involves a high initial cost and there is a possibility of leakage and spills during excavation and transportation. In addition, it is often difficult to determine the exact location of the contamination and to dig in some locations, such as heavily populated areas or under buildings. Exposing the contaminant to the air may lead to other problems, and pumping may just move the contamination around in the aquifers, so it is necessary to monitor the water continuously with test wells.

Excavation and Treatment

Dig out the contaminated earth materials and treat them with a new chemical process that separates the mercury compounds, concentrates them, and mixes them with clay. The clay mixture is baked in a high-temperature kiln where the mercury combines tightly with the clay. The rock-like product does not dissolve in water. These waste "rocks" can then be used in road building since the mercury can no longer dissolve in water and be released into the environment. Building the plant that changes the contaminated earth into waste "rocks" is expensive. There is also some danger that mercury vapors will be released into the air during the heating process.

Electrification

Use an underground electric field to decontaminate the groundwater. The process changes the electric charge on the pollutants, which makes it easier to remove them from the soil. This is less expensive than other options and is effective for the entire area. It is quick and has been safe in all tests to date. However, since it has only been tested in small areas, there is some concern that it might not work in this large area. The direction and power of the electric field may not be as precisely controlled over this large area it has been in the smaller tests.

Shut Down the Water Source

If alternate sources of water cannot be located in the area, the residents of Silver Oaks may have to bring water in from a more distant source. This will possibly involve more cost and certainly involve inconvenience. An alternate source of water may not be readily available. It does not clean up the contaminated groundwater, and will not protect plants and the wildlife from the effects of the mercury.

QUESTIONS

1. What method(s) do you favor? Why?
2. What method(s) do you dislike? Why?

20 The Town Meeting

You have mapped the underground contamination plume and know where the source is located. Now it is time for the community to take action. Science provides evidence, but people make decisions.

GROUP CHALLENGE

Your challenge is to help the town of Silver Oaks decide on the best way to clean up the mercury contamination.

PROCEDURE

Hold a town meeting to decide on the best method for cleaning up the mercury contamination in Silver Oaks. Use the Role Playing Tips and Audience Participation Tips on the following page. Do your best to act your assigned role.

Role players: Don't forget the persuasive power of evidence as you prepare your presentation!

Silver Oaks Beacon

Threat to Planned Shopping Mall

The recent discovery of mercury contamination on the proposed site of the new Silver Oaks shopping mall has halted progress on approval of the site. Local laws do not allow the land to be developed until it is cleaned up. Until the pollution was discovered, the development plans appeared to have the go-ahead. Final approval by the City Council was scheduled for next month. The downtown shopping area can no longer serve the growing population of Silver Oaks and the surrounding area. Citizens surveyed by this reporter were disappointed with the new turn of events. "I was so excited, but now I'm worried that the new mall will never be built," said one young shopper.

ROLE-PLAYING TIPS

1. You are playing the part of a person involved in this issue. How are you involved and what might your opinion be?
2. What reasons might you have for this opinion?
3. Who might not agree with you? (Consider the other roles.)
4. What reasons might be given for the disagreement?
5. How would you respond to opposing arguments?
6. List the key points that you want to make clear in your presentation to the audience.
7. How would you behave at a public meeting? How would you dress? How would you greet people? How would you talk?

You may want to bring props and dress in costume for your part.

It is strongly suggested that you practice your presentation in advance.

AUDIENCE PARTICIPATION TIPS

1. You are playing the role of a member of the audience. How does the issue affect you?
2. Have you already decided what you think should be done?
3. Can someone convince you to change your opinion?
4. What questions do you want answered at the meeting?
5. How will you dress for the meeting?
6. How will you behave at the meeting?

During the meeting, you will be given an index card on which to write questions for members of the panel. The questions can be those you thought of earlier or any that may occur to you during the meeting. Direct your questions to specific members of the panel. Be sure that you get all the information you need to make a thoughtful decision.

Materials

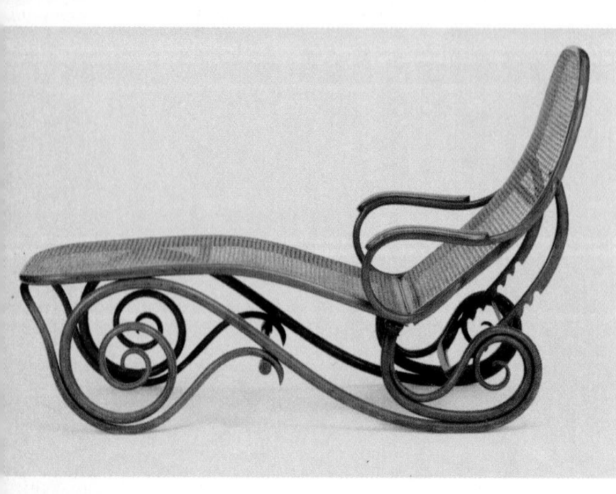

Materials

Materials science deals with the selection, production, and processing of materials for a wide range of uses. Advances in civilizations have taken place because people have learned to craft new materials or to use materials in new combinations and designs.

Materials are used in every aspect of modern life—homes, schools, businesses, and factories, and from industry to recreation to agriculture. In the following activities, you will learn how scientists and engineers select materials for specific purposes. You will also learn about plastics, the newest materials produced since the discovery of metal alloys. Plastics are designer materials. They can be created and designed to be used for specific purposes, such as super glues or even artificial heart valves.

The materials we create can eventually cause problems. When their useful life ends, they become wastes that are discarded into the Earth's air, water, and land. In this unit, you will study about the problems of these "leftovers" that we call garbage. You will also learn about the environmental impact of buying, using, and discarding consumer products.

Materials touch your life in many ways. Just about everything in your world is a result of materials science—from the fillings in your teeth to the plastic pens you use for writing to the personal stereos, microwaves, or other electronic products in your home. Let's turn the page and take a closer look at the world of materials.

Hand Craft

Throwing things away is an ancient—and still common—human activity. Is it something only humans do, or do other kinds of living things also throw away unwanted objects? Maybe it's the throw itself that contains some clues about this behavior. Picking up something, holding it, and then doing something with it (like throwing it) seems so natural that we hardly give the task a thought. But what would tasks that require the use of your hand be like without your thumb?

Try to carry out a variety of activities without using your thumb.

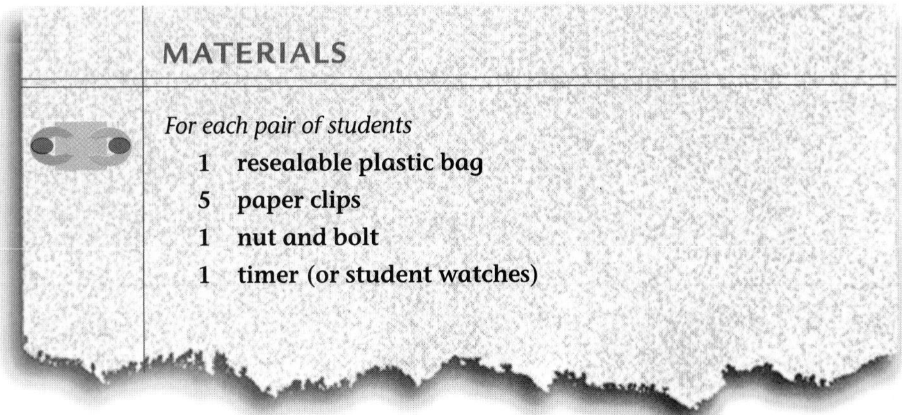

MATERIALS

For each pair of students

- 1 **resealable plastic bag**
- 5 **paper clips**
- 1 **nut and bolt**
- 1 **timer (or student watches)**

PROCEDURE

You will pit your **manipulative** skills against those of your partner, and then pit your team's skills against those of your classmates by playing a game called "Terminal Phalanx." (**Manipulate** means to skillfully manage or move with the hands.) The basic idea of the game is simple. Each team is timed on how long it takes both members to complete a series of tasks both with and without the use of their thumbs. The team with the shortest combined time is the winner. Before beginning, decide which team member will do the tasks first and which one will be the observer. (You will be doing both so it doesn't really matter.) When you are the observer, you must time your opponent and make sure that he or she is not using either thumb during Round 2. Make a data table in your science notebook to record times.

THE TASKS

- Unzip the resealable plastic bag and remove the contents.
- Make a chain with the 5 paper clips by linking them together.
- Thread the nut completely onto the bolt—all the way to the end.
- Put the chain and threaded nut into the bag and zip it tightly.

THE RULES

- Play begins with the sealed plastic bags on the table in front of each opposing player. Each bag must contain 5 paper clips (unlinked) and a nut and bolt (unthreaded). You may not touch the bag until time is started.
- Round 1 allows the use of thumbs. Be sure that you switch roles and reseal the plastic bag so that everyone gets a recorded time.
- Round 2 must be completed without the use of thumbs. When you are the player, you must tuck your thumbs into the palms of your hands and keep them there. If the observer notices that the thumb was used, the player must start that individual task over again. Switch roles.
- Record the times for you and your partner for each round. Compute the total times for your team.

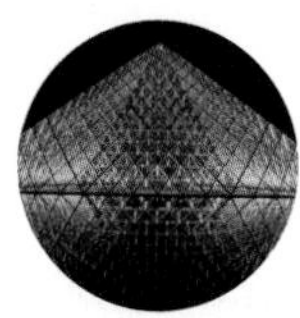

In the Company of Others

Many human behaviors, including using materials, are also demonstrated to some extent by other living organisms. It's likely that early people even learned from watching animals.

CHALLENGE

After completing this reading, think of other examples of complex human activities that are also carried out by animals.

Although we share the planet with many other kinds of living things, it is easy for us to feel different and unique. But can we agree on what is human, and only human? Most likely, you can think of other living organisms that learn, play, fight, have feelings, and communicate. For example, scientists have observed chimpanzees laughing as they tickle or play tricks on one another. Other animals perform elaborate courtship rituals to communicate. The courtship display and ritual dance of male and female cranes is their way of overcoming their natural fear of one another and forming a strong bond. Once a pair mates, they stay partners for life.

Chimpanzees

Human family relationships aren't unique either. Having parents (or sometimes adoptive parents) that care for their young is common among other animals. Ostrich families even set up daycare. After the young ostriches hatch, they gather together in multi-family groups, which are looked after by just one adult.

We aren't even alone in the basic ways that we get our food. Some animals "farm" or tend "livestock." In South America there is a type of ant that "farms." The ants work together carrying pieces of leaves to their underground nest, where they use the leaf pieces as fungus garden beds. The ants "sow" the beds with the spores of a particular quick-growing fungus. These garden beds can be several feet in length. Another type of ant raises aphids. These ants will even fight to protect the aphids from their enemies. The ant colony keeps the aphids near a good supply of food. In exchange for looking after the needs of the aphids, the ants are able to "milk" the aphids for the sweet nectar they produce by gently stroking their backs. Doesn't this seem a lot like dairy farming?

For a long time humans thought that one trait that distinguished them from other living organisms was the ability to use tools to make life easier. But humans are not the only tool-users. The Egyptian vulture picks up a stone in its beak and throws it to break open the thick, hard shell of ostrich eggs. Other birds use plant parts as tools. The woodpecker finch, a bird that lives on the Galapagos Islands, uses cactus spines to pick out insects and grubs from the bark of trees.

Some animals actually adorn themselves with "clothes." The caddisfly larva, which lives in water, weaves itself a tube-shaped silk covering and then sticks bits of gravel, twigs, and leaves onto it. This creates a camouflage fashion statement!

QUESTION

Can you think of any human activities that are not done by animals? Just what is it that sets us apart from other animals?

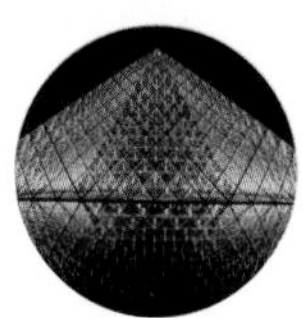

Human Use of Materials

Look at the objects on this page and the next. Some are hundreds or thousands of years old, while others are from our own time. These items are made of different materials.

CHALLENGE

Observe each object carefully and try to determine what materials it contains. Using the number of each picture, work with your group to put the pictures in order by age of the object with the oldest first.

1

2

3

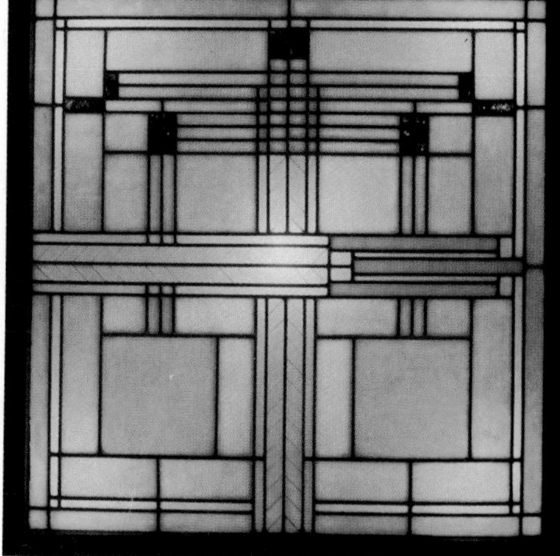

4

5

6

7

8

9

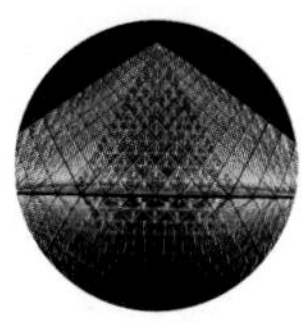

Garbology Survey

Much of what we know about the materials used by people before us is based on a study of what they threw away. What might future generations learn about us by examining the remains of our garbage? Today's landfills are filling up fast. In order to deal effectively with garbage problems, it is important to know not only what the problems are but also which problems the public thinks are important. Since you and your classmates are members of the public, the following survey of your ideas will provide some evidence about what one group of the public thinks.

CHALLENGE

Answer the survey question below in your science notebook. Be sure to give your reasons for your answer. Do you think you have enough evidence to make a decision?

Most of the studies of garbage indicate the following are the contributors to garbage problems. Which one do you think is the greatest cause of garbage problems? Rank these, starting with the one that you think takes up the most space (volume) in landfills.

a. Disposable diapers

b. Food and yard waste

c. Newspapers

d. Large appliances

e. Construction debris

f. All paper

g. Plastic bottles

Who's Fooling Whom?

Studying garbage can tell us many things about garbage and people. In this reading, two students survey their community and get some results that conflict with their observations.

CHALLENGE

Consider the value and limitations of using surveys as a method to obtain evidence.

THE GREAT GARBAGE MYSTERY

Willis and Darnell live in a big city in the northeastern United States. The neighborhood around their school has a problem with garbage. There is a lot of trash on the streets, in vacant lots, and around the school's athletic field. Cans, bottles, and paper make up a lot of the trash. All of these items can be recycled. Every week in their neighborhood, the city's recycling truck picks up cans, bottles, and paper for recycling.

Dumpster, New York City

For science class, Willis and Darnell decide to survey people who live in the area around their school. The survey asks how often people recycle their cans, bottles, and paper. The results of the survey are surprising! Almost everyone who was surveyed says he or she always recycles paper, cans, and bottles.

Seeing all the cans, bottles, and paper trash in their neighborhood, Willis and Darnell wonder whether there is something wrong with their survey. Most people who live in the neighborhood say they recycle these items. So why, the students ask themselves, are there still a lot of cans, bottles, and paper trash scattered around the neighborhood?

QUESTION

What could explain the difference between the students' survey results and the amount of trash that they see in their neighborhood? Try to think of two or three possible explanations.

Using Archaeology to Learn About Today

This reading will help you learn more about The Garbage Project and the science of garbology, including its relationship to archaeology.

CHALLENGE

If researchers from The Garbage Project asked you to tell them what was in your garbage last week, would they find the same discrepancies between your descriptions and the evidence that is described in "The Great Garbage Mystery"?

THE GARBAGE PROJECT

Archaeology tells us many things about the past. It can tell us how people behaved, what they believed, and what they did in their everyday lives. Garbology is a branch of archaeology that looks at people *today* rather than in the past. The Garbage Project studies garbology. The

Municipal garbage, Wallington, New Jersey

Project's scientists have collected evidence about many human behaviors, from what we buy and eat to how we spend our leisure time.

The Garbage Project started at the University of Arizona in the early 1970s. It began from some projects conducted in a college anthropology class. The students were trying to see if there were links between various objects and how people behave. Some students chose to study garbage. The study's results were so interesting that the teacher of the class decided that investigating garbage could give us evidence about our own behavior.

Studies of our garbage often provide evidence that does not agree with what people say they believe or do. For example, in a national survey conducted by the U.S. Department of Agriculture, people said they were eating less junk food and more healthy foods. But data collected by The Garbage Project showed something different. By analyzing packaging and other food wastes from fresh garbage and landfills all over the U.S., the group found that people eat more junk food, sugar, candy, potato chips, bacon, and ice cream than they say they do. The Garbage Project also found that people eat less healthy food, such as cottage cheese, liver, tuna, and vegetable soup than they say they do.

Garbage studies like these cannot tell us whether people are fooling themselves or just want to present a positive image to others. In any case, The Garbage Project studies can help us learn more about nutrition, food waste, and consumer habits than we could from just asking people about these subjects in a survey.

Garbage studies can also tell us about what people buy, read, have as hobbies, and even what they do for fun! As you can imagine, not everyone wants someone looking through their garbage. For this reason, all data collected are confidential. For over 20 years, The Garbage Project studies have challenged some beliefs about human behavior and the overall makeup of our garbage. This kind of information provides a foundation of evidence to help us make good decisions about waste problems and waste management.

For further reading: *Rubbish!* by William Rathje and Cullen Murphy, New York: Harper Collins, 1992.

QUESTIONS

1. What people say and what they do, are not always the same. In your science notebook give an example from The Garbage Project that supports this statement.
2. Why do you think The Garbage Project got these results?
3. Think of a question you could answer by investigating garbage.

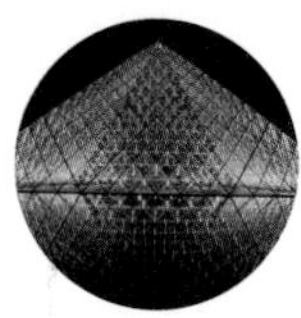

What's in the Landfill?

The chart below compares what the public perceived was the greatest problem in garbage to the actual volume in landfills found by The Garbage Project.

CHALLENGE

Compare the public perception of the greatest problem in garbage to the evidence unearthed by The Garbage Project. How would you explain the differences?

Public Perceptions versus Evidence of Landfill Volume		
Item	**Results of 1990 Roper Poll of the Public**	**Actual Volume in Landfills as Found by The Garbage Project**
Disposable diapers	41%	2%
Plastic bottles	29%	<1%
Large appliances	24%	2%
Newspapers	11%	13%
All paper	6%	40%
Food and yard waste	3%	7%
Construction debris	0%	12%

Hands-on Career Training

Many people follow interesting career pathways. For example, Sheli Smith has been the curator of the Los Angeles Maritime Museum and has participated in other archaeology education projects. Before that, she worked as an archaeologist in charge of the excavation of a ship that was buried in a part of Manhattan built as construction fill. The Garbage Project played an important role in her training for these jobs. In the mid-1970s she worked for The Garbage Project, once telling a *Wall Street Journal* reporter that she sorted garbage "to relax." When she worked as a garbage sorter, Smith appeared on the television show "To Tell the Truth." In this television show, which was very popular in the 1960s and '70s, celebrity panelists tried to guess the profession of guests. Before the show, Sheli Smith had a special manicure. None of the panelists were able to discover her job. One of the panelists said that "No one with nails like that would ever sort garbage."

From: *Rubbish!* by William Rathje and Cullen Murphy, New York: Harper Collins, 1992.

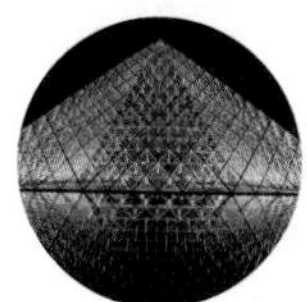

Materials and Dwellings

Look carefully at the pictures of buildings on this page and the next. All of them are dwellings—homes. Try to identify the materials used to make each building.

CHALLENGE

After trying to identify the materials used in each building, decide on the advantages and disadvantages of each material.

From the earliest times, buildings have been made from materials that were easily available. The nature of the material used helped determine how strong the structure could be. It also affected how comfortable the people were inside. The structures we live in are meant to shelter and protect us from the changes in our environment—heat, cold, wind, rain, and snow.

Igloo, Northwest Greenland

House, London

QUESTION

Look at the structures on this page and the one before. How effective would each be in protecting its inhabitants? Relate your answers to the material(s) each is made of.

Sod house, France

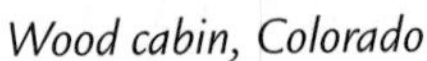

Wood cabin, Colorado

Buffalo tipi, Nebraska

Adobe dwelling, New Mexico

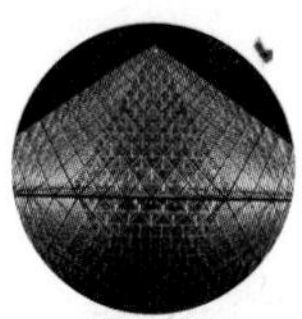

Investigating Properties

You will investigate a wide variety of materials. How can you tell the difference between one material and another? The properties of materials tell us about their similarities and differences.

CHALLENGE

Think about how the properties of different materials help determine the ways in which they can be used.

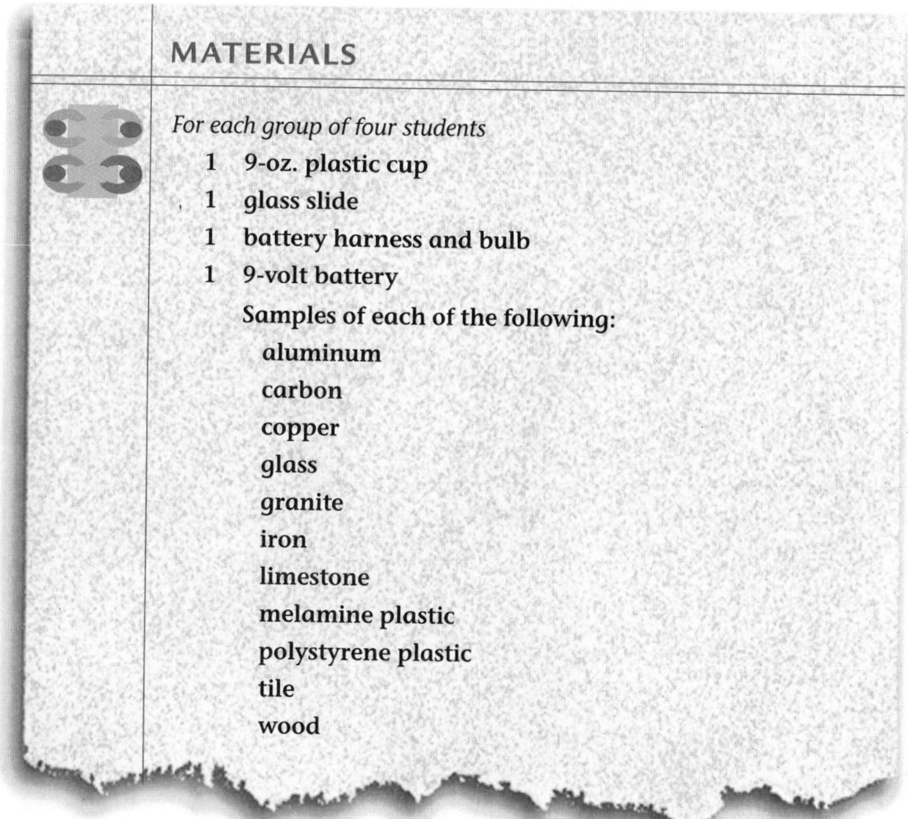

MATERIALS

For each group of four students

- 1 9-oz. plastic cup
- 1 glass slide
- 1 battery harness and bulb
- 1 9-volt battery

Samples of each of the following:

- aluminum
- carbon
- copper
- glass
- granite
- iron
- limestone
- melamine plastic
- polystyrene plastic
- tile
- wood

PROCEDURE

SAFETY NOTE: If a material is not easily flexed when you try to bend it, do not exert extra pressure to break or tear it. Be careful of sharp edges! Wear safety eyewear.

Observe and record the following properties for each of the materials provided: color, light transmission (how much light passes through it), luster (the way light is reflected from it), texture, flexibility, resiliency, hardness, electrical conductivity, and density. Use the information below to help you with the tests for each property. Record your observations in a data table in your science notebook. The more complex tests are described below.

Hardness

To test the hardness of each material, determine if it is harder or softer than glass. Gently press the material across the surface of the glass plate. If a scratch appears that is not easily rubbed away, the material is harder than glass. If no scratch is seen, or if the scratch is easily rubbed away, the material is softer than glass.

Testing for density

Electrical Conductivity

Test the electrical conductivity of each material by attaching the bulb and battery harness assembly to separate ends of the object. If the bulb lights, the material conducts electricity.

Density

To test the density of each material, compare its density to the density of water. Fill the plastic cup half full of water and place the material in the cup. Check to see whether it sinks or floats. Push any material that floats under water with your stir stick and see if it returns to the surface.

If it floats, it is less dense than water; if it sinks, it is more dense than water. As soon as you have tested the material, quickly remove it and dry it.

ANALYSIS

Put the materials you have tested into groups based on their properties. Each group must have one, two, or more properties in common. Record your groupings in your science notebook.

Materials and Human Performance

New materials have improved the way we dress, get to work, cook, and eat. In this reading you will discover how changes in materials have changed the performance of athletes in the pole vault.

After finishing the reading below, think of other examples of new materials that have contributed to improvements in human performance.

OLYMPIC MATERIALS

In 1935, Brutus Hamilton, a track and field coach at the University of California at Berkeley, made some startling predictions. He said that under perfect conditions a man might someday run a mile in just over 4 minutes, high jump almost 7 feet, and pole vault over 15 feet—but those would be the limits. People laughed at his predictions because they seemed so impossible.

Well, we now know that it is possible to run a mile in well under 4 minutes. We also know that Coach Hamilton's predictions in the high jump and pole vault have been passed. The following bar graphs show the Olympic records over the years. The improvement is highlighted for easy reference.

High Jump Olympic Records

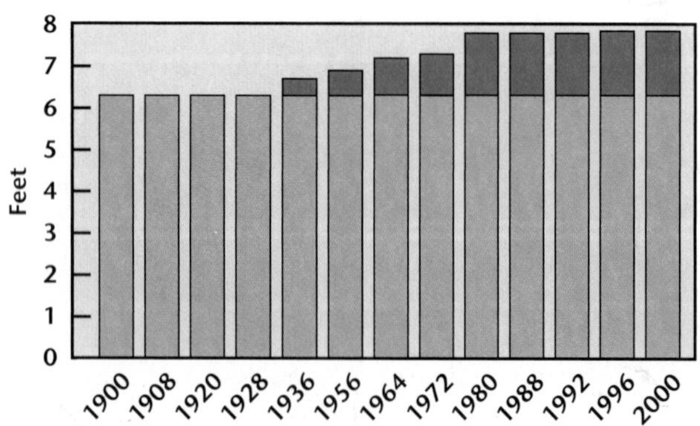

Pole Vault Olympic Records

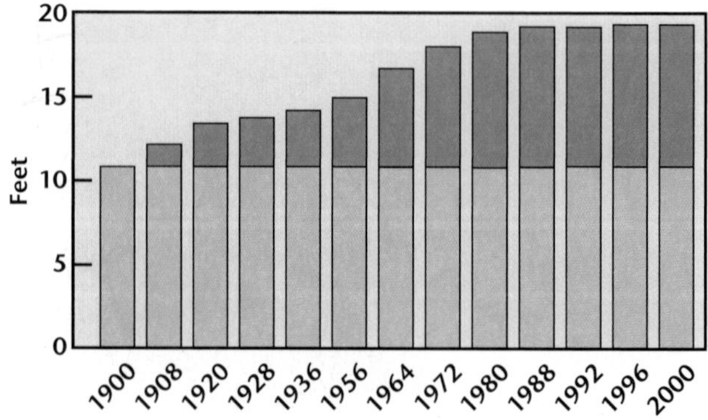

Both sports have benefited from improved training practices. Synthetic track surfaces and lighter nylon running shoes have also helped. In addition, pole vaulters now use poles made of new composite materials, including Kevlar and boron. The poles used today are very strong and flexible, a vast improvement over the rigid metal poles used in the early 1900s, the bamboo poles used in the 1920s and 1930s, and the plain fiberglass poles used until recently. The athletes must also appreciate the soft foam pads they land on instead of the old pits full of sawdust that were used in the past.

Pole vaulting—then and now

Pole vaulting has changed in other ways from the early days of the sport. Computer models now project the maximum height possible given the speed of the runner at the instant when the pole is "planted" for takeoff, the kind of grip used, the twist of the vaulter as the pole flexes back from its bent position, and other variables. This gives the athletes information they can use to try to improve performance. In addition, pole vaulting rules do not require that all poles be the same. Different materials are favored by different athletes. Some athletes even treat the pole differently to improve their grip—some athletes use sticky tape or spray adhesives, and others smear a glue-like substance on the end of the pole. Athletes are constantly on the lookout for ways to use materials to improve their performance.

QUESTION

Examine the Pole Vault Olympic Records and High Jump Olympic Records graphs closely. Which graph shows the greater increase? Can you think of any reason for this difference?

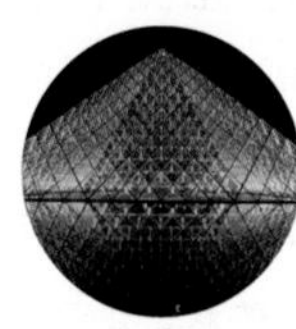

Conductivity of Common Materials

Another property that affects how we use a material is its ability to conduct heat. The rate at which heat energy is transferred by a material is called its **thermal conductivity.**

CHALLENGE

Using the chart on the next page, compare the thermal conductivity of different materials. What properties do good thermal conductors have in common?

QUESTIONS

1. Which two materials are the poorest conductors of heat? The best?
2. What can you generally say about the relationship between how much heat a material conducts and its density?
3. If all other factors are held constant, which one of the three little pigs would be the coolest in his house on a hot, calm summer day—the one that made his house of straw, sticks, or bricks?

Conductivity of Common Materials		
	Thermal Conductivity*	**Density (g/cm^3)**
Aluminum	164.0	2.7
Common brick	4.2	1.75
Concrete	2.5	1.4
Copper	278.0	8.92
Fiberboard	0.35	0.24
Fiberglass, blanket	0.2	0.05
Glass, window	5.5	2.5
Gypsum, board	1.0	0.8
Ice	12.5	0.9
Iron	55.0	7.86
Limestone	11.0	2.5
Marble	15.0	2.6
Paper	0.7	0.9
Plaster	4.2	1.8
Plastics, foamed	0.2	0.2
Plastics, solid	1.1	1.2
Straw	0.2	0.3
Wood, balsa	0.3	0.16
Wood, oak	1.0	0.7
Wood, pine	0.7	0.5

*Btu/(hr.) ($ft.^2$) (°F/ft.)

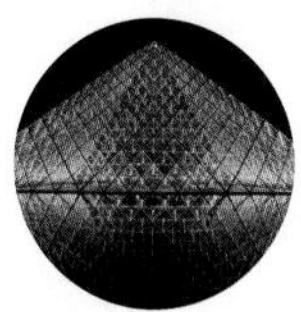

Try and Try Again

This reading describes Thomas Edison's efforts to perfect the electric light bulb. His persistence is illustrated by the many materials he tried and the amount of time he spent trying to find the right design.

CHALLENGE

Your challenge is to identify the variables that Thomas Edison kept constant and the one he changed as he tried to build a better light bulb.

TOO SMALL, TOO RED, TOO HOT

"Too small, too red, too hot," said Thomas Edison in frustration one day in his lab over 100 years ago. Day after day, for over two years, he and his assistants experimented with thousands of different kinds of filaments (the part of a light bulb that actually gives off the light), trying to find just the right design. They were either so thin that they broke, or they gave off light that was too red, or they heated up so much that they melted. Even though the electric light was not his invention, Edison had a vision of a world lit at night by inexpensive, long-lasting electric lights. He even said that the day would come when only the rich would be able to afford candles. The problem was finding just the right materials to do the job. His early failure led to success in the end.

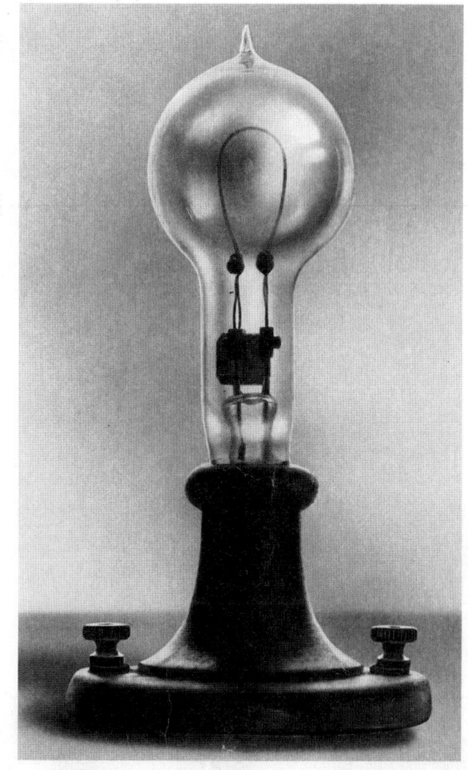

Early light bulb

The basic idea was to send enough electricity through a conducting material to make it heat up and glow. If the material was shaped into a very thin, long strand, more heat and light would be given off. Edison knew that metals were good conductors of electricity, but he also knew that most would melt long before they gave off much light. He also wanted to produce a bright white light, not a dim red glow. The metal platinum held promise, though, so Edison announced that not only had he perfected the electric light but also that he had found an almost limitless source of platinum—a rare and expensive metal. He said that there were great amounts of platinum in the black sand of the Pacific Coast beaches. Neither claim turned out to be true. The platinum lights made at his workshop in New Jersey were very costly and burned out in just a few hours.

For many months he worked on designs that were sensitive to temperature. When the filament came close to melting, the electricity would automatically shut off. They were complicated, expensive, and never worked well. The lights would often melt, and sometimes they would even explode, sending out a shower of sparks and glass. Those that did not explode would constantly flicker off and on. Certainly these were not going to replace the candle or gas lamp!

Modern lamp

Edison then came up with the idea of using carbon (the black soot that he noticed on the inside of his gas lamp) for the filament of the light. Like metal, carbon conducts electricity, although not as easily. But the main advantage was that it didn't melt at high temperatures. Edison collected some of the carbon residue from a gas lamp and coated a thin cotton string with it. He then baked the coated string in a hot oven to harden it. To his dismay, it was extremely brittle and would break even before he could put it in a glass bulb. He had to put the carbonized string in a glass bulb, remove most of the air to form a vacuum, and then seal it shut so that it would work well.

Edison in 1878, 31 years old

For the next several years, Edison searched for the perfect material to coat with carbon—one that could be made into very thin strands, but was not as brittle as the cotton string when it was hardened. He and his assistants tried cork, cardboard, wood shavings, coconut hair, leather, pasta, onion rinds, and even their own hair. Each experimental filament would take hours to make. It was not uncommon for Edison and his assistants to work 12 to 15 hours a day, sometimes even sleeping at the lab so that they could begin early the next morning.

In the end, after testing over 6,000 different kinds of materials, Edison chose a particular type of bamboo for the light bulb filament. The idea for using bamboo came to him when he picked up a Japanese fan sitting on his desk. Today, we take the light bulb for granted, but its development is one of the great stories in the history of materials science.

Modern bulbs use a very thin tungsten wire for the filament. Tungsten was chosen because it is easy to form into a tight coil, increasing the light output, and it has a high melting point. It can reach 4,500°F (2,500°C) and not melt.

QUESTIONS

1. Edison once said invention is 99% perspiration and 1% inspiration. What did he mean?
2. How important was finding the right material for the filament of the electric light bulb? Explain your answer.

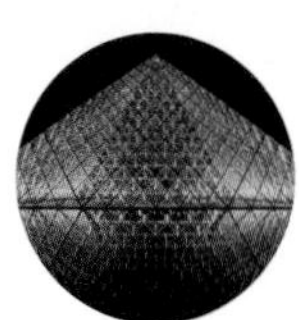

Investigating Corrosion

One important property of any material is how well it will last in the environment where it will be used. For example, metals sometimes corrode, which affects their usefulness. In this activity you will investigate the effect of a number of factors on the corrosion of iron and other metals. You will consider the environmental trade-offs involved in various approaches to the prevention of corrosion.

CHALLENGE

Determine the advantages and disadvantages, including environmental effects, of different approaches to preventing the corrosion of iron.

In the junkyard

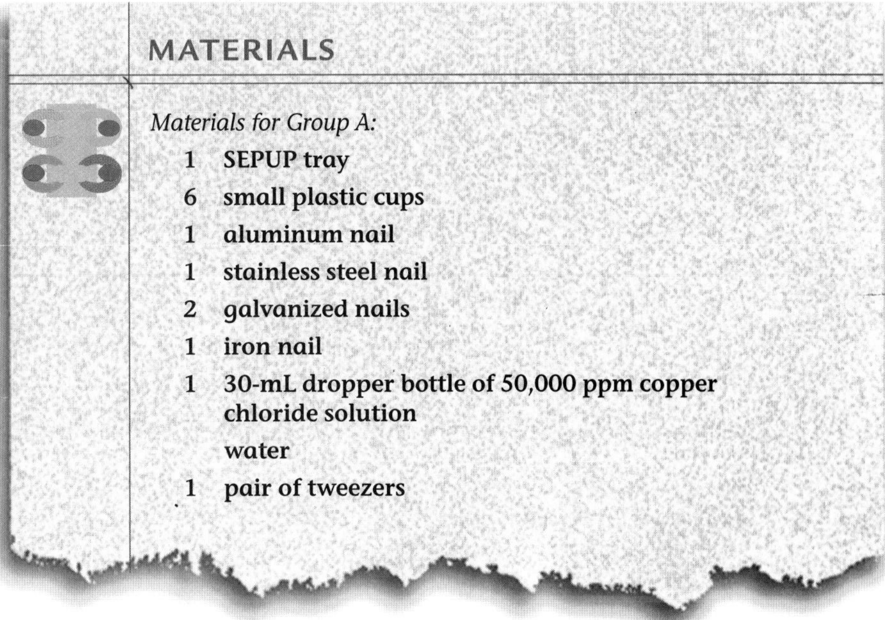

MATERIALS

Materials for Group A:

1 SEPUP tray
6 small plastic cups
1 aluminum nail
1 stainless steel nail
2 galvanized nails
1 iron nail
1 30-mL dropper bottle of 50,000 ppm copper chloride solution
water
1 pair of tweezers

MATERIALS

Materials for Group B:

1 SEPUP tray
6 small plastic cups
1 aluminum nail
1 stainless steel nail
2 galvanized nails
1 iron nail
1 30-mL dropper bottle of 50,000 ppm copper chloride solution
water
1 packet of salt
1 graduated cylinder
1 9-oz. clear plastic cup
1 stir stick
1 pair of tweezers

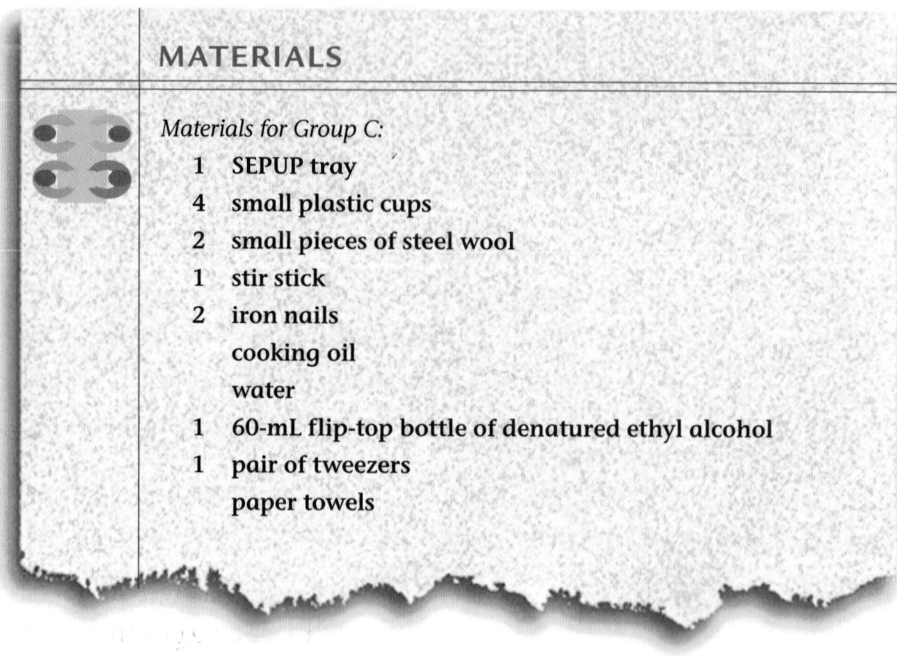

MATERIALS

Materials for Group C:

1 SEPUP tray
4 small plastic cups
2 small pieces of steel wool
1 stir stick
2 iron nails
cooking oil
water
1 60-mL flip-top bottle of denatured ethyl alcohol
1 pair of tweezers
paper towels

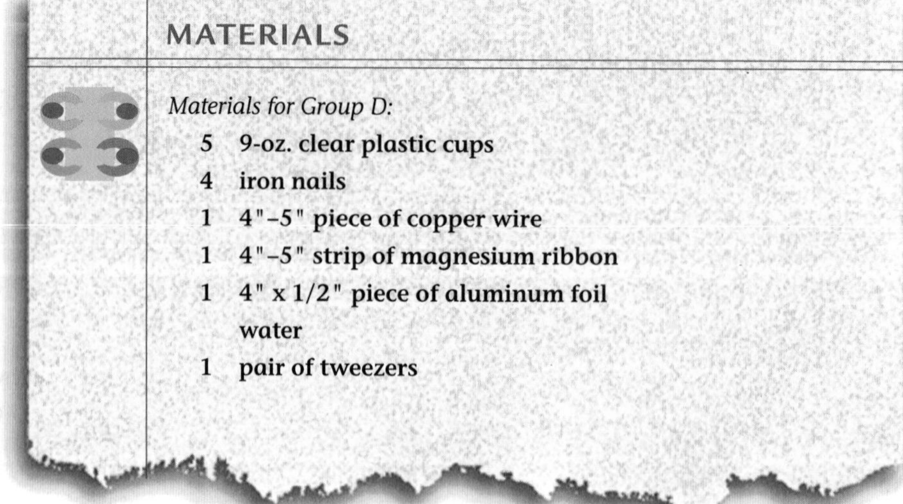

MATERIALS

Materials for Group D:

5 9-oz. clear plastic cups
4 iron nails
1 4"–5" piece of copper wire
1 4"–5" strip of magnesium ribbon
1 4" x 1/2" piece of aluminum foil
water
1 pair of tweezers

In this activity there are four different investigations, identified as A, B, C, and D. Your teacher will assign you to one of the investigations.

PROCEDURE A

SAFETY NOTE: Wear protective eyewear. Some people may have an allergic reaction to the copper chloride solution, resulting in itching and redness in the affected area for a short time. Wash any exposed area with water for 2–3 minutes.

Testing Corrosion of Different Kinds of Nails in Fresh Water

1. Use the SEPUP tray and five small plastic cups. Insert one plastic cup into each of the large cups in the tray, labeled A–E.
2. Place the aluminum, stainless steel, and iron nails in Cups A, B, and C, respectively. Place one galvanized nail in Cup D and one in Cup E.
3. Add enough water to fill the plastic cups in Cups A–D about three-quarters full.
4. Add enough copper chloride solution to fill the plastic cup in Cup E about three-quarters full.
5. Wait 15 minutes, and then record your observations in a data table like the one shown on page B-37.
6. A galvanized nail is an iron nail that is coated with zinc. What does the copper chloride solution appear to do to the nail? Record your ideas about what has happened to the galvanized nail in your science notebook.

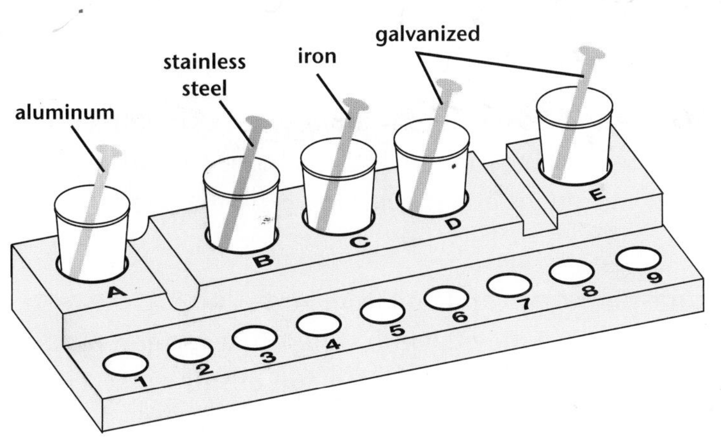

7. Now remove the plastic cup and the nail from large Cup E. Use the tweezers to hold the nail. Rinse the nail with water and put it in a clean plastic cup three-quarters full of water. Put this cup in large Cup E of your SEPUP tray.
8. Return the used copper chloride solution to the large bottle provided by your teacher.

9. Wait 15 minutes, and then record your observations in a data table in your science notebook.

10. Follow your teacher's instructions for storing the samples. You will observe your results and present them to the class in the next class session.

PROCEDURE B

SAFETY NOTE: Wear protective eyewear. Some people may have an allergic reaction to the copper chloride solution, resulting in itching and redness in the affected area for a short time. Wash any exposed area with water for 2–3 minutes.

Testing Corrosion of Different Kinds of Nails in Salt Water

1. Prepare a solution of salt water by adding one packet of salt to a 9-oz. plastic cup. Use the graduated cylinder to add 60-mL of water to the salt. Stir thoroughly.

2. Use the SEPUP tray and 5 small plastic cups. Place one plastic cup into each of the large cups of the tray, labeled A–E.

3. Place the aluminum, stainless steel, and iron nails in Cups A, B, and C, respectively. Place one galvanized nail in Cup D and one in Cup E.

4. Use the salt water you prepared in Step 1 to fill each of the cups in Cups A–D about three-quarters full. Put the rest of the salt water aside. You will need it for Step 8.

5. Add enough copper chloride solution to fill the plastic cup in Cup E about three-quarters full.

6. Wait 15 minutes, and then record your observations in your science notebook using a data table like the one on the next page.

7. A galvanized nail is an iron nail that is coated with zinc. What does the copper chloride solution seem to do to the nail? In your science notebook, record your ideas about what has happened to the galvanized nail.

8. Next, remove the plastic cup and nail from large Cup E. Use tweezers to hold the nail. Rinse the nail with water. Take a clean plastic tasting cup and fill it about three-quarters full of salt water. Place the nail in the cup. Put this cup in large Cup E of your SEPUP tray.

Data table for Procedures A and B

Corrosion of Nails in Fresh and Salt Water

Cup	Type of Nail	Exposed To 15 Minutes	Observations After 2nd Day	Observations on
A	Aluminum			
B	Stainless steel			
C	Iron			
D	Galvanized iron			
E	Galvanized iron	copper chloride	1st observation 2nd observation	

9. Return the used copper chloride solution to the large bottle provided by your teacher.
10. Wait 15 minutes, and then record your observations in your data table.
11. Follow your teacher's instructions for storing the samples. You will observe your results and present them to the class in the next class session.

PROCEDURE C

SAFETY NOTE: Wear protective eyewear. Denatured ethyl alcohol is toxic and flammable.

Testing the Effect of Oil Coating and Size of Surface Area

1. Take two small pieces of steel wool. Place one in a small plastic cup and add enough denatured ethyl alcohol to just cover the steel wool. Allow it to sit for 2 minutes. Then, pour the alcohol back into the bottle and, using tweezers to hold the steel wool, rinse the steel wool and plastic cup with water several times. Place the plastic cup

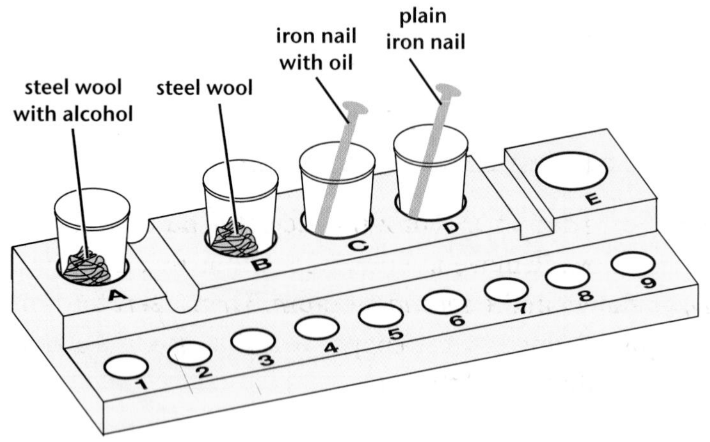

containing the steel wool into large Cup A of a SEPUP tray.

2. Place the other small piece of steel wool into another plastic cup and set the cup in large Cup B of the SEPUP tray.

3. Take the two iron nails. Coat one nail with some cooking oil. To do this, place several drops of cooking oil on a paper towel. Wipe the oil-soaked towel over the nail so that it is completely coated. Be sure to rub the oil all over the nail. It must be completely covered with cooking oil. Place the uncoated nail and the oil-coated nail in two separate plastic cups. Put one of these cups in large Cup C and the other in large Cup D of the tray.

4. Fill each of the plastic cups about three-quarters full of water.

5. Wait 15 minutes, and then record your observations in your science notebook in a data table like the one below. Title your table: Effects of Oils and Surface Area.

6. Follow your teacher's instructions for storing the samples. You will observe your results and present them to the class in the next class session.

Data table for Procedures C and D

Title: ____________________

Cup	Object Tested	Exposed To	Observations After 15 Minutes	Observations on 2nd Day
A				
B				
C				
D				

PROCEDURE D

SAFETY NOTE: Wear protective eyewear.

Testing the Effect of Contact with Other Metals

1. Take the copper wire and wrap it around one of the nails, as shown in the diagram below. Remove the nail from the wire. Stretch the wire slightly, as shown in the second diagram. Then insert the nail back into the spiral formed by the copper wire. The stretched wire should fit quite tightly around the nail.

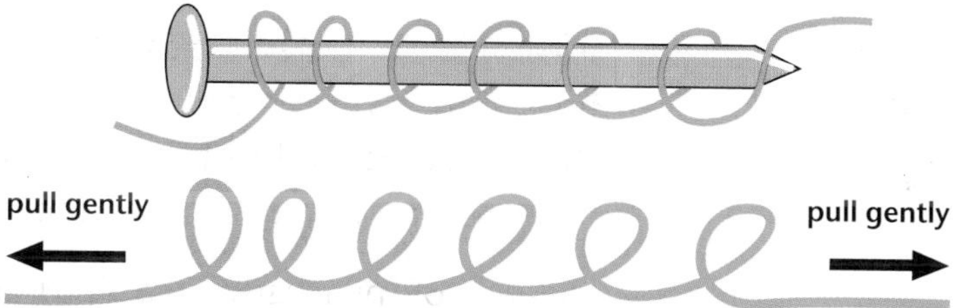

2. Repeat the same process using the magnesium ribbon and a nail. Be extremely careful when you stretch the magnesium ribbon. It breaks very easily. Do not pull very hard on it.
3. Take the third nail and wrap it in aluminum foil. Roll the foil up tightly and then wrap it around the nail.
4. The fourth nail will not be wrapped in another metal. Take four 9-oz. plastic cups. Put each of the four nails into one of the cups. Next, fill each cup about one-third full of water.
5. Wait 15 minutes, and then observe the nails. Record your observations in your science notebook in a data table like the one shown on the previous page. Title your table: Effects of Other Metals.
6. Follow your teacher's instructions for storing the samples. You will observe your results and present them to the class in the next class session.

Iron and Steel

Iron and steel are widely-used materials. This reading describes how the development of the ability to use metals, especially iron and steel, has changed our lives.

CHALLENGE

Think about the many ways that you use iron, steel and other metals in your life.

People have been using metals for thousands of years. Gold and silver have long been used to make ornaments, but they are too soft to use for weapons or tools. The Bronze Age began about 3500 B.C. with the discovery of methods for making bronze by combining copper and tin. Bronze was much harder than any individual metal then available, but because copper was rare and costly, most people continued to use bone and stone tools. It wasn't until methods for producing iron were discovered that a metal became widely available as a material.

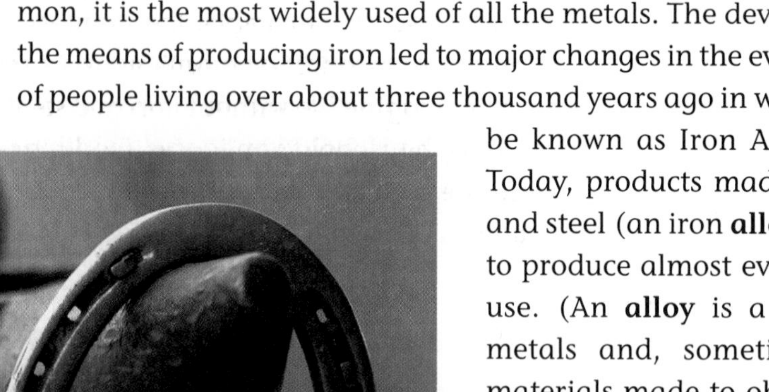

Iron is the fourth most abundant element in the Earth's crust and is found on every continent. Because it is strong, inexpensive, and common, it is the most widely used of all the metals. The development of the means of producing iron led to major changes in the everyday lives of people living over about three thousand years ago in what came to be known as Iron Age cultures. Today, products made from iron and steel (an iron **alloy**) are used to produce almost everything we use. (An **alloy** is a mixture of metals and, sometimes, other materials made to obtain metals with certain characteristics.)

When horses were the primary means of long-distance transportation, horseshoes were one of the most important objects made of iron. Why would a horse need an iron shoe?

People first began to use iron they found in meteorites in the period from 6000 to 4000 B.C. They chipped and hammered the meteorites to make weapons, tools, and ornaments. As people learned to harness fire and make it burn hotter by using bellows to push more air into it, the temperatures they could obtain rose high enough to extract iron from rusty-looking ores. In China, India, and Turkey, from 2500 to 1400 B.C., iron was produced by smelting ores. When the Hittite Empire in Turkey fell and the Hittite tribes scattered, the knowledge of iron smelting from ore spread.

About 1000 B.C. the Iron Age began. Iron smelting was independently discovered in Africa in approximately 500 B.C., near Lake Victoria in an area that is now Burundi and Rwanda. Iron was used for tools, weapons, and eventually the armor of the Middle Ages. By A.D. 1000, iron was used to make tools for farming, such as horseshoes and plows. The increased production of tools and machines made possible by the use of iron led to the Industrial Revolution and to great changes in farming as farms became more mechanized. In the 1860s, methods for producing large quantities of steel led to a further revolution in the production of iron-containing products. This permitted the mass production of many products by steel machinery and tools.

Iron is generally used in the form of steel—iron combined with other metals or carbon. The addition of these other elements allows the iron and steel industry to produce an endless variety of products to meet special needs. For example, the stainless steel used to make tableware is produced by adding chromium and carbon, in specific proportions, to iron. Imagine life without iron and steel products. Automobiles, skyscrapers, household appliances, food cans, and the machines that produce the clothes you wear and the food you eat are just some of the products that would not be available!

QUESTIONS

1. In ancient times, what was the first source of iron?
2. What are alloys? How are they used?
3. What differences would there be in your daily life if there were no iron or steel?

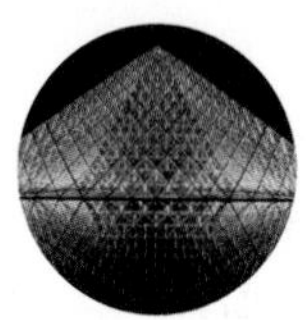

Protecting Liberty

Now that you have performed your corrosion experiments, you will read about the efforts made to control the corrosion process on a very important national monument, the Statue of Liberty located in New York Harbor.

CHALLENGE

Apply your knowledge about the corrosion of metals to solve a practical problem.

She stands 151 feet tall, weighs 280 tons, and, on a clear day, is visible for 42 miles. Yet the Statue of Liberty is more than just a monumental figure. She has welcomed generations of immigrants and visitors to New York's harbor and has come to symbolize the United States of America to millions of people around the world. Protecting Liberty from the effects of corrosion, like protecting liberty itself, is not something we can turn our backs on.

The Statue of Liberty

When the French sculptor Frederic August Bartholdi designed the Statue of Liberty, he knew that he couldn't let the iron support structure inside the statue touch the copper skin on the outside. Why not? The skin is made of 300 thin copper plates joined together by some 300,000 copper rivets. The problem is galvanic corrosion. This occurs when two different metals—in this case, iron and copper—come in contact in the presence of an electrolyte (a solution that conducts electricity by the movement of ions). In galvanic corrosion, one of the two metals corrodes faster than the other. In this case, iron corrodes up to 1,000 times faster than copper!

Scaffolding helped workers repair the statue

To prevent galvanic corrosion, Bartholdi and his structural engineer Gustave Eiffel (who later designed the Eiffel Tower in Paris) put shellac and asbestos strips between the iron ribs and copper skin. This prevented the metals from touching so corrosion did not occur. But as Liberty's metal parts expanded and contracted with changing temperatures and the push of the wind, the insulator strips soon wore away, and the incompatible metals made contact. Seawater collected in the statue, and corrosion began quickly. Also, the rust building up between the iron and copper created a pressure great enough to pop out thousands of rivets, leaving holes in the skin for even more salt water to seep through.

After standing for almost 100 years in New York Harbor, the corrosion was so bad that the Statue of Liberty had to be closed for a massive rebuilding project. The iron bars holding up the copper skin had lost, in some cases, up to two-thirds of their original size.

As part of the rebuilding, all the flat iron bars in the skin-support system —more than 1,300 bars totaling 35,000 pounds—were replaced with stainless steel substitutes. Stainless steel is an alloy of iron, chromium, nickel, and molybdenum that is corrosion resistant. The old copper skin and new stainless steel support bars are now lined with Teflon—a slippery polymer that is an electrical insulator—to provide added protection against galvanic corrosion and to reduce problems caused when the statue expands and contracts in response to heat and cold.

Adapted with permission from *ChemMatters*, April 1985, © 1985 American Chemical Society.

QUESTIONS

1. What is galvanic corrosion? How does it affect the Statue of Liberty?
2. How can you prevent or slow down galvanic corrosion?
3. In some parts of the United States, roads are salted in the winter to melt ice. As you would expect, this increases the rate of corrosion of automobiles. Based on the investigations done in your class, what are some possible methods that could be used to prevent corrosion as a result of the salt? Which method would you recommend? Explain your choice.

6 Bag it! Paper or Plastic?

Comparing Bags

In this activity you will compare the properties of two materials commonly used in grocery bags—paper and plastic.

CHALLENGE

Design an ideal bag for your local supermarket. Identify what qualities (properties) your bag would have and why they are important.

PROCEDURE

1. Work with your group of four students to list as many advantages and disadvantages as you can for the use of both paper and plastic materials for grocery bags. Make a table to organize your ideas.
2. Paper bags and plastic bags do not magically appear when needed and disappear when we finish using them. Where do they come from? Where do they end up? The entire manufacturing process, from the resources used to make a product to the final fate of the product, is called the product life cycle. In your science notebook, copy the product life cycle diagrams below and write how you think having a diagram of the cycles helps you decide which bag to choose. What else do you want to know?

Life Cycle of a Bag

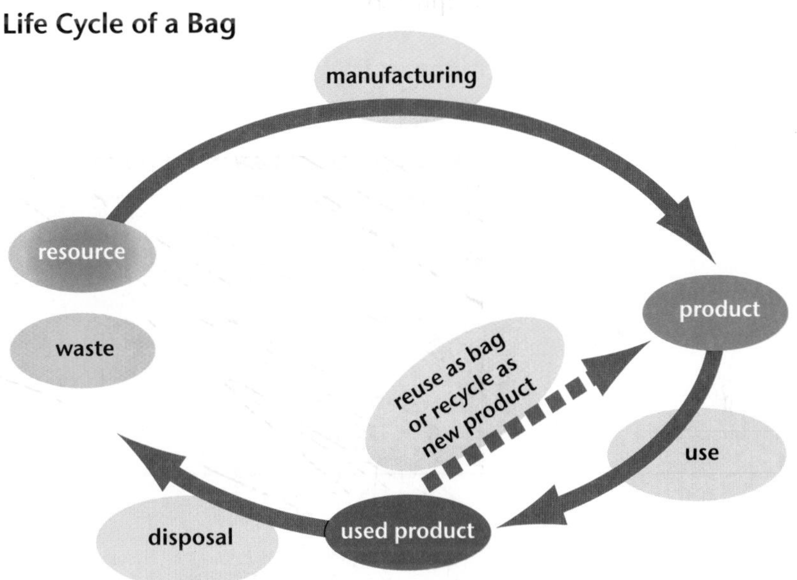

QUESTION

What do you think about the choices given in the cartoon—to "deplete oil reserves or cut down a forest?" Are these the only choices? Explain your answer.

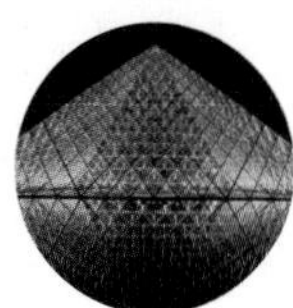

Physical Properties of Plastics

You have examined the physical properties of a variety of materials. Now you will examine the physical properties of plastics.

CHALLENGE

Analyze the physical properties of four known plastics. Think about how these properties will affect how each plastic can be used.

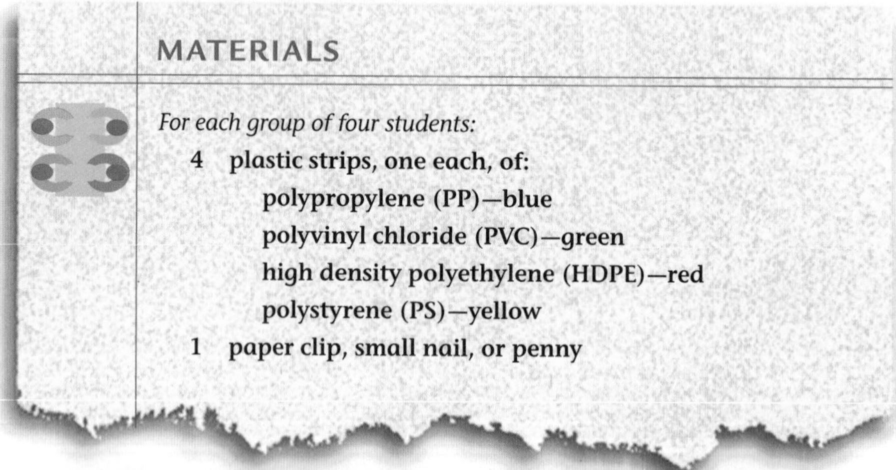

MATERIALS

For each group of four students:

4 plastic strips, one each, of:
- polypropylene (PP)—blue
- polyvinyl chloride (PVC)—green
- high density polyethylene (HDPE)—red
- polystyrene (PS)—yellow

1 paper clip, small nail, or penny

Plastic items come in all shapes and colors.

PROCEDURE

Each member of your group should test a different plastic strip by using the tests outlined below. Record your observations in your science notebook in a data table like the one below.

1. Flexibility and Crease Color

Gently bend your strip back and forth to observe its flexibility (ability to bend) and the color of the crease that is produced. Record your observations.

2. Hardness

The hardness of a material is its resistance to being scratched. Test for hardness by using the end of a paper clip, nail, or penny to see if it will scratch the plastic. Use only gentle pressure; do not gouge the sample! Record your observations.

3. Acetone and Heat

Watch your teacher demonstrate the effects of acetone and heat on each of the four types of plastic. Record your observations.

Testing Plastic Strips

	Student Tests				Teacher Demonstrations	
Plastic Name	Color	Flexibility (excellent, good, fair, poor)	Crease Color	Hardness (scratches)	Effect of Acetone	Effect of Heat
Polypropylene (PP)	blue					
Polyvinyl chloride (PVC)	green					
High density polyethlene (HDPE)	red					
Polystyrene (PS)	yellow					

Density

Density is a physical property of all materials. It is the mass of a substance per unit of volume. Relative density can be determined by comparing the weights of the same volume of different materials.

Determine the relative densities of the four plastics by comparing whether they float or sink in various liquids.

MATERIALS

For each group of four students:

- 1 package salt
- 4 plastic graduated vials, with caps, labeled 1, 2, 3 and 4
- tap water
- 1 9-oz. plastic cup
- 1 60-mL bottle of denatured ethyl alcohol
- 1 dropper
- 4 small, square plastic pieces, one each, of:
  - polypropylene—blue
  - polyvinyl chloride—green
  - high density polyethylene—red
  - polystyrene—yellow
- 1 pair of tweezers

PROCEDURE

SAFETY NOTE: Denatured ethyl alcohol is toxic and flammable. Wear safety eyewear.

1. Label the four vials: 1, 2, 3, and 4.
2. Measure 10 mL of denatured ethyl alcohol into Vial 1. Place the cap on the vial.
3. Measure 5 mL of denatured ethyl alcohol and 5 mL of tap water into Vial 2. Place the cap on the vial. Shake well to mix.
4. Measure 10 mL of tap water into Vial 3. Place the cap on the vial.
5. Measure 10 mL of tap water into Vial 4. Pour in one package of salt. Place the cap on the vial and shake well to mix.
6. Each student in the group should test all four plastic squares in one of the four solutions. Prepare a data table to record your group's observations. The first student should place all four plastic squares in Vial 1, cap the vial, and shake gently to eliminate the effect of air bubbles or surface tension.
7. Everyone in the group should observe the experiment and record his or her observations in a data table.
8. The person who tested the squares should now remove them by using tweezers and placing them on a paper towel. Blot the squares with the paper towel and give them to another member of your group to test in Vial 2. Continue this until the plastics have been tested in all four solutions. Be sure your data table is complete.

ANALYSIS

1. Which plastic was the least dense? What is your evidence?
2. Which plastic is the most dense? What is your evidence?
3. Place the four plastics on this scale of specific density.

Density Scale (Water=1)

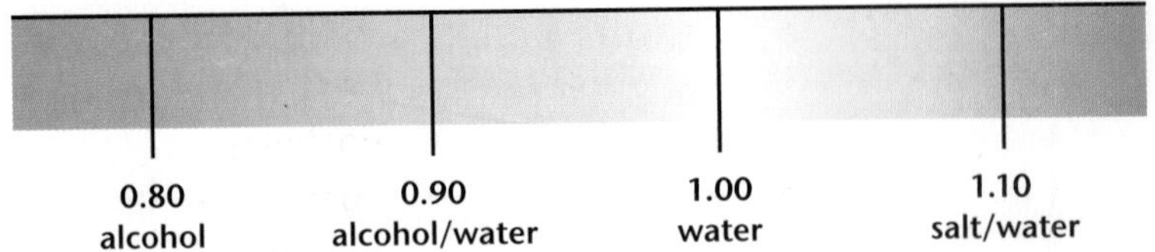

4. Which plastics were the most flexible? Name a product in which this is a desirable quality and explain.
5. Which plastic was the least flexible? Name a product in which this is a desirable quality and explain.
6. Imagine you are a toy designer. Why would you want to know about the crease color of a plastic you might use in a toy?
7. Which plastic(s) scratched? How will this property affect their use?
8. Which plastic(s) were affected by acetone?
9. You are the design engineer for containers of nail polish remover (high in acetone). Of these four plastics, which one(s) would you consider using? Why?
10. Which plastics could be used for "dishwasher safe" food-storage containers? Explain.

A LIST OF COMMON POLYMER PRODUCTS

The following is a list of common items which contain polymers. In your science notebook, list any of these items that you have encountered in the past seven days.

ashtrays
attaché cases
audiotapes
awnings
baby bottles
baby rattles
backpack
balloons
Band-Aids®
basketballs
beverage boxes
bicycle and automobile paint
bicycle handgrips
boats
bubble gum
bubble pack
butane lighters
buttons
cameras
car batteries
carpets
caulking
ceiling light covers

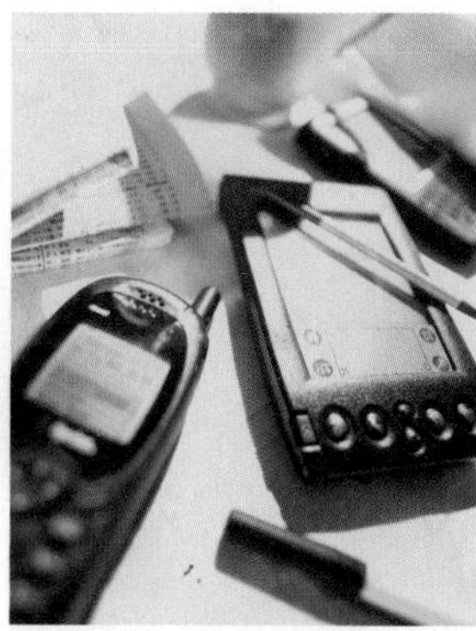

Personal electronics

cellophane tape
chair seats
checkers
clarinets
cleats
coasters
coffee mugs
coffee stirrers
combs
compact discs (CDs)
computer discs
computers
contact lenses
credit cards
curtains
cushions
dashboards
decorative fruit
desks
dishpans
disposable diapers
disposable razors
dust brushes
earphones
egg cartons
electrical tape
epoxy glue
erasers
exercise mats
extension cords
false eyelashes
false teeth
fan belts
fast-food containers
fishing bobbers
fishing line

flags
flea collars
floor mats
floor tiles
floor waxes
flowerpots
flutes
foam cups
foam rubber pillows
folding doors
food storage containers
food wraps
football helmets
football pads
footballs
furniture polishes
galoshes
garbage bags
gears
glasses
guitar strings
hair curlers
hair dryers
hair sprays
handles
hang gliders
hearing aids
hockey pucks
house paint
ice chests
ice cube trays
index tabs
insulated foam cups
insulation
jewelry
knapsacks

Sponges

lawn chairs
life preservers
life rafts
lip gloss
lipstick tubes
luggage
lunch boxes
lunch trays
mannequins
margarine tubs
measuring tapes
metallic balloons
microwave cookware
milk jugs
model cars
model planes
mops
motor oil bottles
motorcycle helmets
movie film
overhead projectors
pacifiers
paint brushes
panty hose
parachutes
particle board
patio screens

(continued on next page)

pencil cases
pens
photographic film
photographs
pianos
picture frames
pillowcases
ping-pong balls
plastic dishes
plastic flowers
plastic glasses and cups
plastic knives, forks, and spoons
playing cards
plywood
portable radios
protractors
puppets
racquetballs
raincoats
recorders
reflectors
refrigerators
rope
rubber bands
rubber duckies
rubber gloves
rubber soles
rubber tubing
rulers

safety glasses
sailboats
sails
sandals
sandwich bags
school desks
shampoo bottles
shoe boxes
shoe polish
shoestring tips
shoestrings
shower doors
shrink wrap
shuttlecocks
skateboard wheels
slides
snack food bags
sneakers
snorkels
soda bottles
sofas
sponges
store signs
street signs
sun visors
sunglasses
supermarket meat trays
surfboards
sweater boxes
sweaters

swim fins
synthetic fabrics for clothing
tabletops
tackle boxes
tape recorders
Teflon® cookware
telephones
tennis balls
tennis racquets and strings
tent pegs
tents
test tube brushes
test tube racks
Thermos® bottles
thread
thread spools
tires
toilet seats
toothbrushes
toothpaste tubes
transparencies
trash cans
trophies
typewriter cases
typewriter ribbons
umbrellas
Velcro®
venetian blinds

Tennis ball

videotapes
vinyl siding
vinyl tops
vinyl wall coverings
vitamin capsules
volleyballs and nets
waders
wallets
watch faces
water pipes
welcome mats
wet suits
whistles
wigs
windbreakers
windshield wipers
windshields
yarn
zippers

Egg cartons

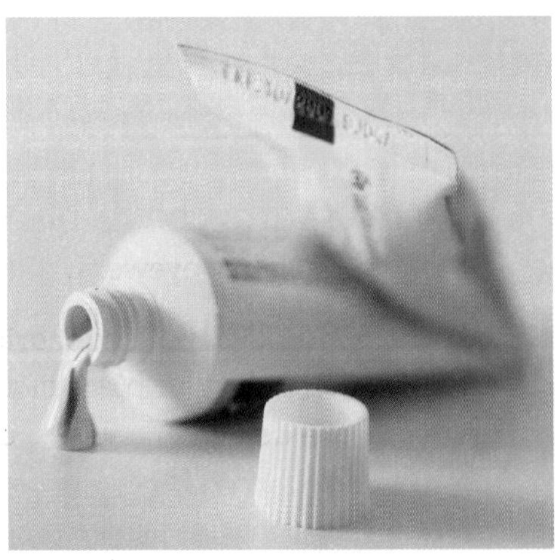

Tooth paste tube

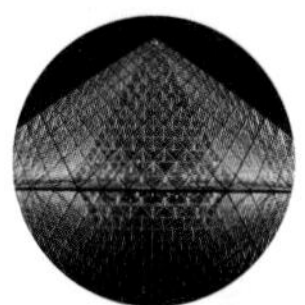

Identifying Unknown Polymers

In this activity you have learned about the physical properties of plastics. Now, you will use the data you have gathered to help you identify some unknown plastics.

CHALLENGE

Identify the unknown plastic sample your teacher provides.

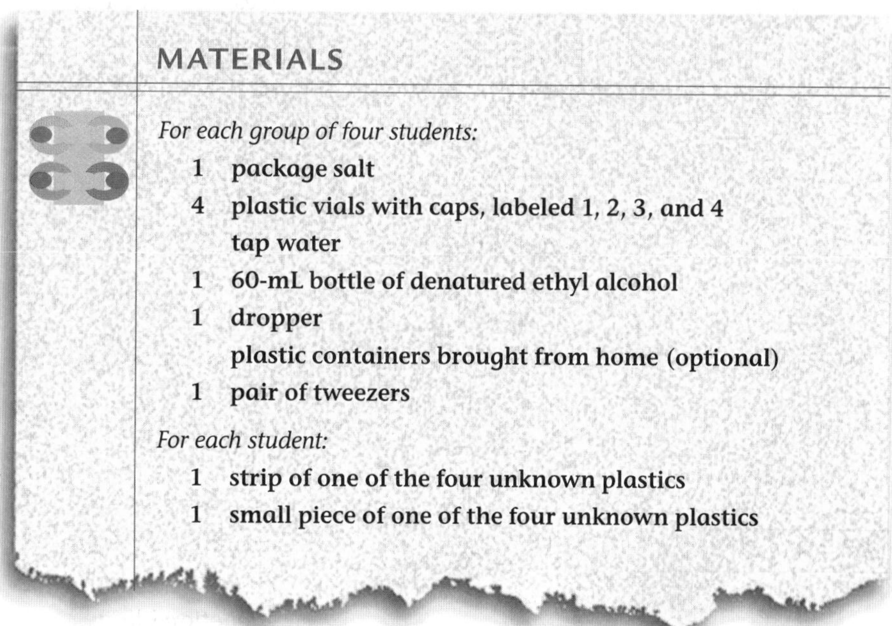

MATERIALS

For each group of four students:

1 package salt
4 plastic vials with caps, labeled 1, 2, 3, and 4
tap water
1 60-mL bottle of denatured ethyl alcohol
1 dropper
plastic containers brought from home (optional)
1 pair of tweezers

For each student:

1 strip of one of the four unknown plastics
1 small piece of one of the four unknown plastics

PROCEDURE

SAFETY NOTE: Denatured ethyl alcohol is toxic and flammable. Wear safety eyewear.

Although you are sharing materials, you will work independently and identify your piece of plastic. Your group should prepare the vials for the density test as you did before.

Testing the Unknown Plastic Sample

1. Using the strip and small square of plastic your teacher gives you, perform whatever tests are necessary to identify the plastic. It is one of the plastics you have studied, but it has a different color. Record the procedures you follow in your investigation and your observations on a separate piece of paper. Think about how you will organize the data you collect so that they are clear.

2. Ask your teacher to demonstrate the effects of acetone and heat on these plastics. Record your observations.

ANALYSIS

1. Use what you have learned about the properties of the four plastics to help identify this unknown piece of plastic. Defend your decision by using your data, and explain how you know that your unknown is the plastic you have chosen, and not the other three.

2. Which of the four plastics from class that you have tested would be the best choice for the following applications? Give evidence from your investigations to support your decision.

   a. Thin film shrink-wrap material

   b. Hinges on a plastic ice chest

   c. Squeeze bottles

   d. Ocean buoys

Testing Plastics from Home

1. Note the source (toy, kitchen tool, etc.) of the small sample of plastic material you brought from home. Your teacher will help you cut up the sample into smaller pieces to make it easier to perform the tests. Record your observations in your science notebook. Think about how you will organize the data you collect so that they are clear.

2. Explain your identification as you did for Analysis Question 1 above.

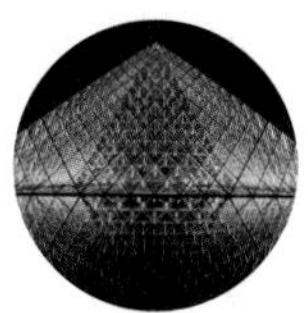

Cross-linking a Polymer

In the previous activities you investigated the properties of common plastics. The plastics you examined are a type of polymer, a substance whose molecules are made out of long chains of smaller molecules. Now you will become a chemical engineer and cross-link the polyvinyl acetate polymer found in white glue.

CHALLENGE

Investigate the properties of the cross-linked polymerized material you create from white glue and sodium borate and suggest possible uses for it.

MATERIALS

For each group of four students:

1 **120-mL bottle of white glue**
1 **30-mL bottle of 4% sodium borate solution**

For each pair of students:

1 **9-oz. clear plastic cup half filled with water**
1 **stir stick**
2 **graduated cups**
paper towels

PROCEDURE

SAFETY NOTE: Be sure to wear protective eyewear. Immediately clean up any spilled materials with water.

1. Set up your data table.
2. Pour 10 mL of white glue into one of the graduated cups. List the properties of the glue in your science notebook.
3. Pour 2.5 mL of the sodium borate solution into the other graduated cup. List the properties of the sodium borate solution in your data table.
4. While stirring with a stir stick, add the sodium borate to the white glue. Closely observe what happens, including any evidence of a temperature change. Continue to stir until nothing further happens.
5. Place the polymerized material in the cup of water to wash off any excess white glue. Remove your polymer from the water and put it on a paper towel. Gently squeeze the material inside the paper towel to absorb the excess water. Note: Avoid getting your polymer on the table, on the floor, or on your clothes.
6. Observe the properties of the new material. Make a list of them. Focus on the following categories:
   a. Stickiness: How does it stick to different kinds of surfaces?
   b. Bounciness: How well does it bounce?
   c. Stretchiness: How well does it pull apart when pulled both quickly and very slowly?
7. Wash your hands and lab equipment thoroughly. Your teacher will instruct you on what to do with the polymer.

ANALYSIS

1. How is the material you made different from the white glue?
2. The synthesis of a polymer involves a chemical change. What evidence do you have that a chemical change occurred?
3. What could you use this new material for?
4. What physical properties would you like to change in the material? Why?

Designing a Material to Meet a Need

Plastics are rarely used in their pure form. Other materials are usually added to polymers in order to change their physical or chemical properties. This is how plastics are designed for specific uses. Adding materials in this way is called compounding.

CHALLENGE

Experiment with compounding the polymer you created previously to make it more suitable for a particular use.

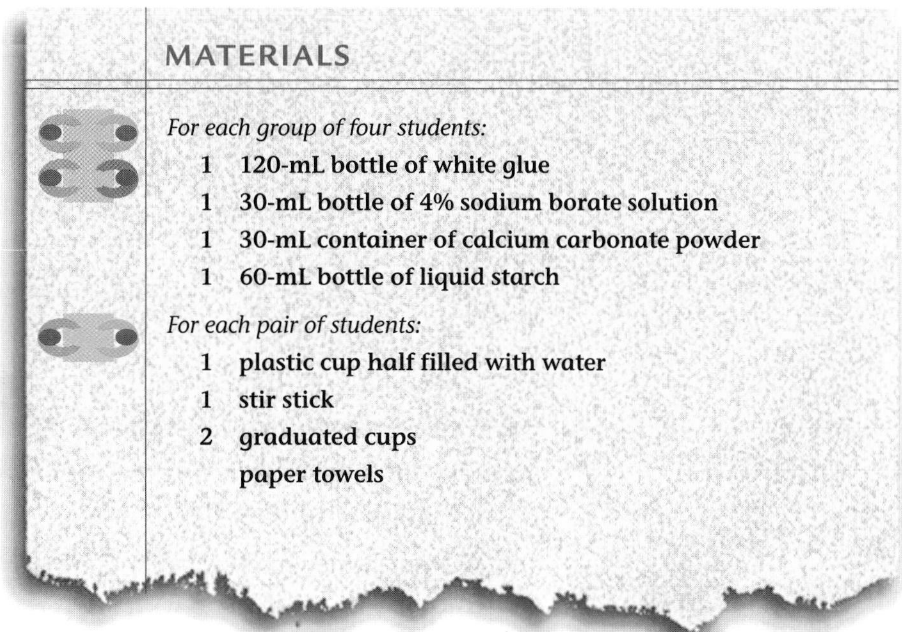

MATERIALS

For each group of four students:

1 **120-mL bottle of white glue**
1 **30-mL bottle of 4% sodium borate solution**
1 **30-mL container of calcium carbonate powder**
1 **60-mL bottle of liquid starch**

For each pair of students:

1 **plastic cup half filled with water**
1 **stir stick**
2 **graduated cups**
paper towels

PROCEDURE

Part One: Creating a Compounded Polymer

1. Work in pairs to do the investigation but set up your own data tables.
2. Add 5 cc (cc, or cubic centimeters, are used to measure powders) of calcium carbonate powder to one of the graduated cups. Then add 5 mL of liquid starch to the same cup and stir. You should now have a total of 10 mL of material in the cup.

3. Now, to this same cup, add 10 mL of white glue and stir. Observe the properties of the mixture. List them in your data table.

4. Pour 2.5 mL of the sodium borate solution into the other graduated container.

5. While stirring, add the sodium borate to the glue mixture. Stir until nothing further happens. Then rinse the new polymer, pat it dry, and observe its properties. List them in your data table.

ANALYSIS

Since atoms and molecules are too small to see, we use indirect clues to tell if a chemical reaction has taken place. Some clues are: color changes, temperature changes, bubbles of gas seen, or change in properties. How is this compounded polymer different from the polymer you made from white glue alone?

Part Two: Investigating How the Ingredients (Reactants) Affect a Compounded Polymer

1. After discussing your results within your group, try the same experiment again. However, this time leave out either the starch or the calcium carbonate powder.

2. Write up your experiment. Prepare to discuss the similarities and differences of this new compounded polymer with other student groups in class. If desired, let your compounded polymer stand overnight and observe any changes.

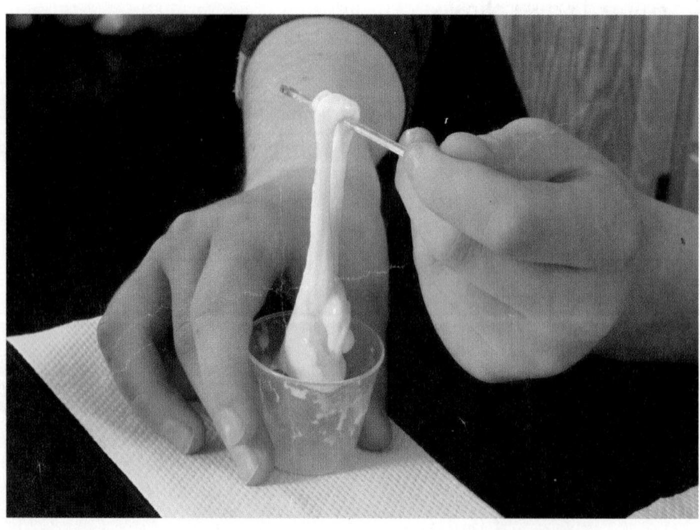

9 Paper Clip Polymers

Making Models of Polymers

In the last activity you synthesized and compounded a polymer. Its properties were changed by the materials you added to it. To help explain what we cannot see taking place, scientists use models. In this activity you will construct paper clip models to help you understand the behavior of the molecules in the substances you combined to make your complex polymers. Keep in mind that a scientific model does not have to look like the real thing—it just has to act like it in an important way.

CHALLENGE

Construct molecular models of polymer molecules that help explain some of the physical properties of your white glue polymers.

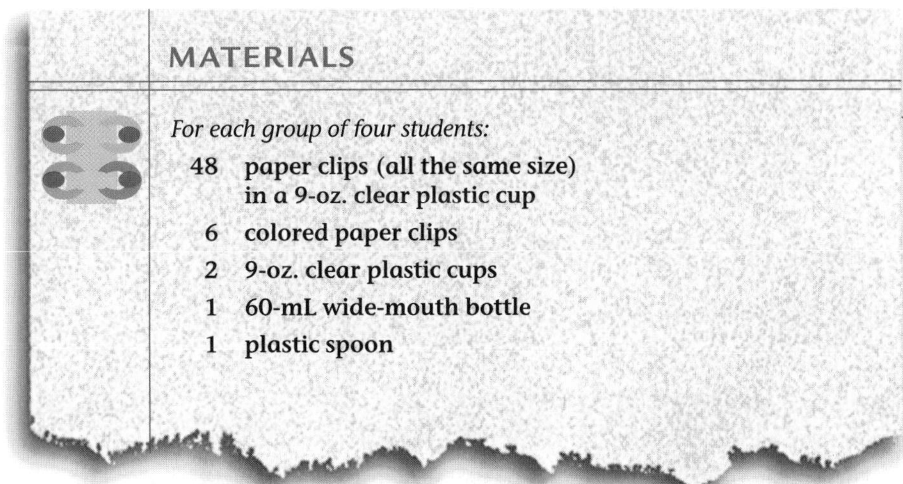

MATERIALS

For each group of four students:

- 48 paper clips (all the same size) in a 9-oz. clear plastic cup
- 6 colored paper clips
- 2 9-oz. clear plastic cups
- 1 60-mL wide-mouth bottle
- 1 plastic spoon

PROCEDURE

Part One: Forming Polymers

1. Put 24 unconnected paper clips into the bottle. These represent monomers. Slowly pour them from the bottle into one of your plastic cups. You may need to shake the bottle to help the clips move. Repeat this two or three times. Describe how quickly the clips come out of the bottle. Record your observations in your science notebook in a data table like the one on the opposite page. Use the row labeled "Monomer" to record your observations.

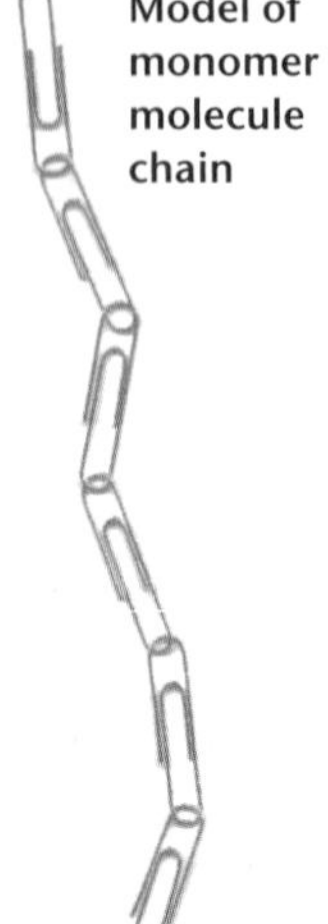

Model of monomer molecule chain

2. Each member of your group should link 6 paper clips together to form a simple chain. Each clip represents an individual monomer molecule.

3. Link all four chains together to make a single chain of 24 paper clips. You have just made a polymer molecule! This chain represents the polyvinyl acetate (found in white glue) solution you began with in the last activity. (You would need thousands of paper clip monomers to make a realistic paper clip polymer. Such a model would be hundreds of times larger than the one you've just made.)

4. Put the polymer you just made in the bottle. Leave one clip hanging outside. Pour the chain into a plastic cup two or three times. Record your observations in the row labeled "Polymer."

5. Now put the polymer molecule into one of your plastic cups and the 24 separate monomer molecules into the other cup. Stir each with the spoon. Record your observations in the row labeled "Stirrability."

6. Remove the polymer chain from the cup. Pull on one end. Describe and record your observations in your data table.

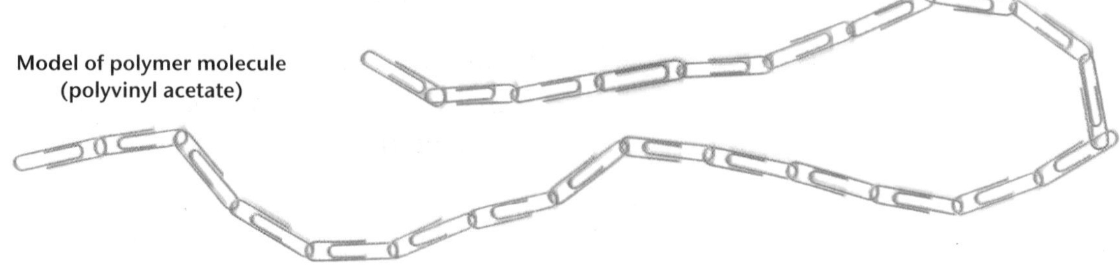

Model of polymer molecule (polyvinyl acetate)

Comparing Model Monomer and Polymer Molecules

	Pourability	Stirrability	Pulling
Monomer			
Polymer			
Cross-linked polymer			

PROCEDURE

Part Two: Cross-linking Polymers

1. Separate the paper clip chain into four equal parts. Each part represents individual polymer molecules. Place the chains in four lines parallel to each other. Use two colored paper clips to make connections between the first and second chain. This is called cross-linking. In the last activity you cross-linked white glue to make a bouncy polymer. The sodium borate acted as a cross-linker, just as the colored paper clips do in your model polymer.

2. Repeat Step 1, using two colored clips between the second and the third chains and two between the third and fourth chains. Now you have a highly cross-linked polymer.

Pouring paper-clip polymers

3. Carefully put your cross-linked polymer into the bottle, leaving one clip hanging out. Pour the polymer into a plastic cup. You may need to work at this. Record your observations in the column labeled "Pourability" for the cross-linked polymer.

4. Now try stirring your cross-linked polymer molecule in the plastic cup. Record your observations in the column labeled "Stirrability."

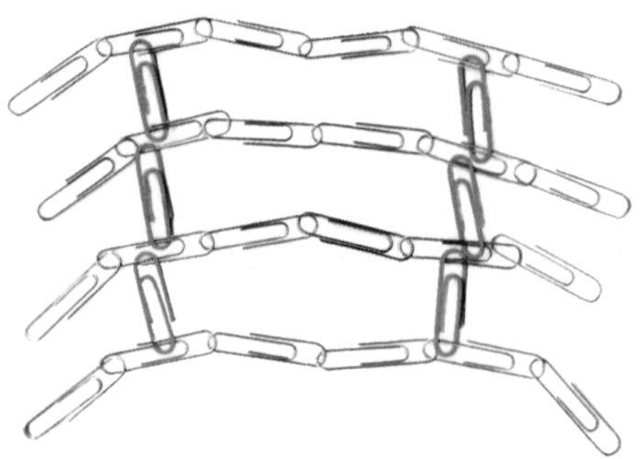

Model of cross-linked polymer

5. Pull on one of the paper clips in your model. Describe what happens to the rest of the paper clips.

6. Separate all of the paper clips and return them to the place designated by your teacher.

ANALYSIS

1. If each paper clip represents a single monomer molecule, how many links or bonds must a molecule be able to make to form a chain like the one in Step 3 in Part One of the Procedure?

2. Explain what causes the difference(s) in pourability.

3. Explain the reasons for any differences you observed in the stirrability of the 24 separate molecules (monomers) and the polymer chain.

4. Would you expect polymer molecules that are connected in long chains to have different physical properties from single monomer molecules? Why?

5. The scientific term that describes how easily a liquid flows is called **viscosity.** Highly viscous liquids are thick, while liquids of low viscosity are runny or thin. How would the viscosity of the monomer compare to the viscosity of the polymer?

6. Write a sentence to describe what happens to the properties of a chemical substance as more and more molecules of that substance are polymerized (forming longer chain molecules).

7. How is viscosity affected by cross-linking? Explain. (Recall what happened to the white glue when you added the sodium borate solution in the last activity.)

EARLY POLYMERS

As the bowling ball rumbled toward the pins, the crowd of spectators covered their eyes and ears. It was utterly quiet. Then, a deafening explosion rocked the room. Once again, all the pins had been destroyed along with the ball. The spectators were elated.

This could have been a true story had not the earliest plastic materials been changed and other new plastics invented.

Plastics have been called "the first new materials to be made in 3,000 years," since the age of the discovery of metals. The earliest plastics were made from natural polymers in materials such as wood and cotton. In England around 1860, Alexander Parkes used cellulose fibers extracted from wood to make the first crude plastic. However, its quality was poor, and it took a contest in America to perfect the new plastic material.

John Wesley Hyatt was aware of the large reward offered to replace the costly ivory used for billiard balls. In 1870, he developed a superior, easily molded product. He won the contest and patented the new product, he called celluloid. It soon became the basis for the earliest movie films.

Early film was made of celluloid

Celluloid was clear and flexible and could be molded easily, but it was very flammable. Projectionists in theaters showing celluloid film kept a large bucket of sand nearby to smother potential flames. Sometimes celluloid billiard balls would smoke and make small explosions as they collided. Even false teeth were made from celluloid. One unlucky smoker needed to have a fire extinguished in his mouth! Today, ping-pong balls are the only product still made from celluloid.

Sorting celluloid ping-pong balls, 1930s

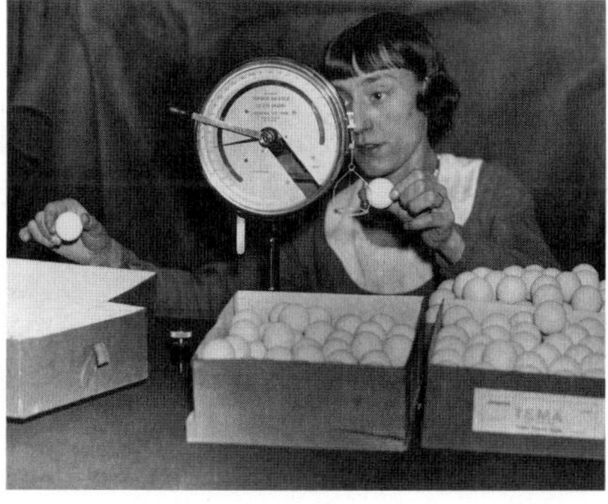

Following the discovery of petroleum in the United States in 1859, plastics came to be made from crude oil. These plastic were synthesized from crude oil, so they are called synthetic plastics. The first such synthetic plastic was made by American chemist Leo Bakeland in 1907. Bakelite is still used today as an electrical insulating material.

QUESTION

1. How did the physical and chemical properties of celluloid limit its early use?

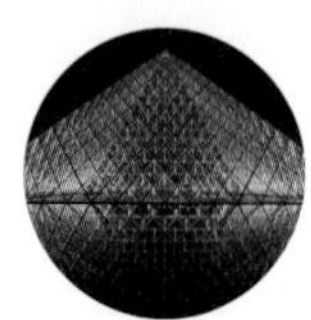

The Big Stretch-Off

You have observed how the linking and cross-linking of molecules affects the properties of a substance. You will now investigate the property of plastic bags that allows them to stretch in different directions. You will then use the paper clip model from the last activity to explain this behavior.

CHALLENGE

Determine whether the direction in which the lines run in a plastic bag affects how the bag stretches. Use the paper clip model to explain this ability to stretch.

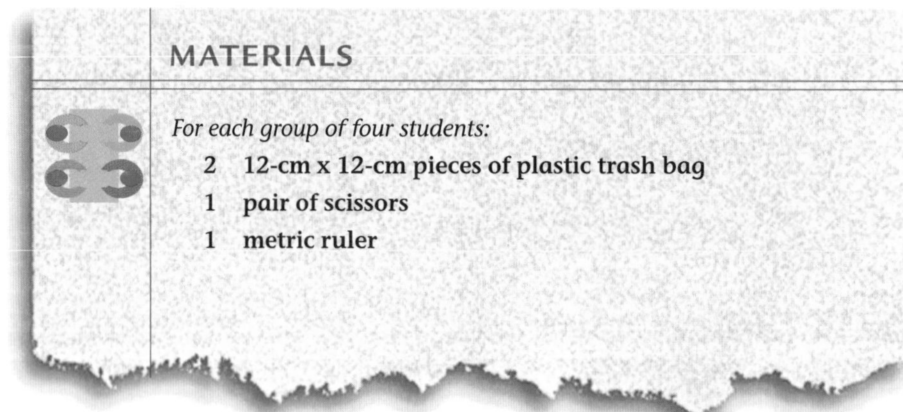

MATERIALS

For each group of four students:

- 2 12-cm x 12-cm pieces of plastic trash bag
- 1 pair of scissors
- 1 metric ruler

PROCEDURE

1. You have two squares cut from a plastic trash bag. Hold them up to the light. Carefully observe! Do you see lines running parallel to each other? You should see lines running in one direction.

2. Next you will stretch a strip along the lines. Have one member of your group take one of the squares and cut it into six 2-cm-wide strips. Cut the strips along the lines so that the lines run from one end of the strip to the other (see the diagram below). Give one strip to each member of your group. Save the other two strips for repeating this step, if necessary.

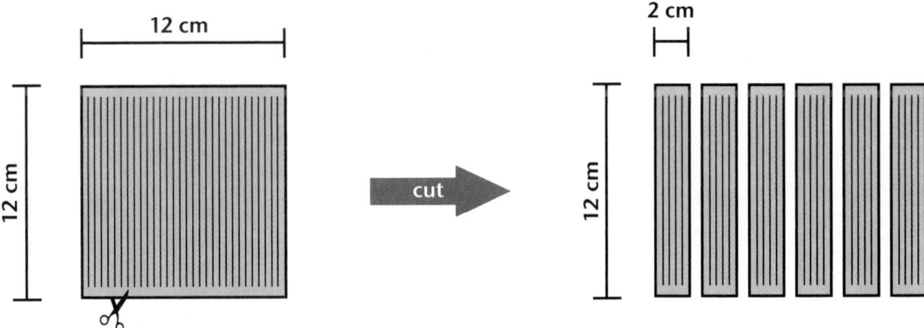

3. Grasp a strip in both hands so that each end extends over thumb and forefinger. Pull it gently. As you pull, look through the strip. What do you see happening? Use your ruler to measure how much each strip stretches before breaking. Record your observations.

4. Next you will stretch a strip against the lines. Have a member of your group cut the other square into six 2-cm-wide strips the other way, so that you cut across the lines in the strip (see the diagram below). Give one strip to each member of your group. Save the other two strips for repeating this step, if necessary.

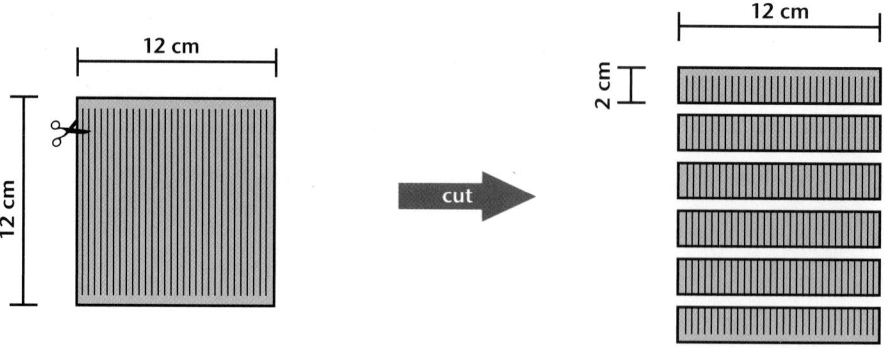

5. Grasp a strip in both hands so that each end extends over thumb and forefinger. Pull it gently. As you pull, look through the strip. What do you see happening? Use your ruler to measure how much each strip stretches before breaking. Describe how the plastic changed as it stretched before breaking. Record your observations.

ANALYSIS

1. In your group, which strips stretched the farthest—the ones cut along the lines or the ones cut across the lines? Do the other groups agree with your group?

2. Describe any changes in the appearance of the plastic that your group observed as the strips were stretched along the lines and against them.

3. Describe how the polymer model could explain the stretching behavior of the plastic bag that you just investigated. It might help you to think about the polymer modeling we did in class with the paper clips. You should try to come up with a hypothesis for how the arrangement of the polymer molecules in the bag could cause the stretching results you observed. Describe your model by drawing a picture and using words to explain it.

Which Packing for Mike's Games?

Imagine that you are the head of the shipping department for a company called Mike's Games. Your company needs a packing material for its new video games. The suppliers of two packing materials have sent you samples and information about their materials along with some samples. The president of the company wants you to perform tests to compare the two materials and to make a recommendation about which polymer the company should use to pack the video games for shipping to customers.

CHALLENGE

Read the background information and advertisements on the following pages, and then test the two sample polymers to decide which material the company should use.

Plastic packing materials come in all shapes and sizes.

TWO KINDS OF FOAM

Polystyrene Foam: Background Information

Polystyrene was first synthesized in Germany in the early 1930s. To make the foam product, small spherical beads of polystyrene are mixed with steam and pentane, a material found in gasoline in small amounts. This causes the beads to expand and soften, allowing the pentane to penetrate inside. The beads containing the pentane are treated with steam again, and then they are injected into a mold where a final application of steam is applied. By this time, the spheres have expanded to nearly 27 times their original volume. The cooled product is extracted from the mold to become a cup or other foamed product. From 1 kilogram (2.2 pounds) of polystyrene beads, over 44,000 cups can be manufactured.

Plastic foam is used for packing items safely for shipping.

In the past, substances called CFCs (chlorofluorocarbons) were used to expand the beads. It was discovered that these destroy the Earth's protective ozone layer, which screens out ultraviolet light. The use of CFCs to expand foam products has been discontinued in the United States and is being phased out worldwide.

Polystyrene Foam: Advertisement

Enviro-Pac

Pack all your products with our new environmentally friendly Enviro-Pac loose fill. It's made from a clean, nontoxic, and lightweight plastic (foamed polystyrene) that provides excellent cushioning for fragile items during shipment. It has greater crush strength than any other filler.

This new foam material contains no ozone-destroying chlorofluorocarbons (CFCs) that can damage the Earth's ozone layer. Instead, it contains a substitute (pentane) that the EPA views as environmentally safe.

Enviro-Pac is superior to newspaper and other less resilient materials. It forms no harmful leachates in landfills because it is chemically inert—it doesn't react or break down. And it can be used in waste-to-energy incinerators (incinerators that use waste as fuel to produce energy) as fuel that contains more energy than coal.

Don't be misled by products that claim to be superior. For cost, ease of handling, moisture resistance, and customer satisfaction, choose Enviro-Pac for all your shipping needs.

Cornstarch Foam: Background Information

An alternate product that is made from a special kind of starch is called Eco-Foam. This is not ordinary cornstarch. It comes from a type of corn grown specifically to produce it. The starch is processed in a pressure cooker with a small amount of synthetic polymer and heated to form a gel-like material with a consistency like gummy candies. When the pressure is released, the gel quickly foams to form the pieces that will be later used as packaging filler. When the foam is wet, the starch and binder quickly dissolve, and the material melts back into a gel.

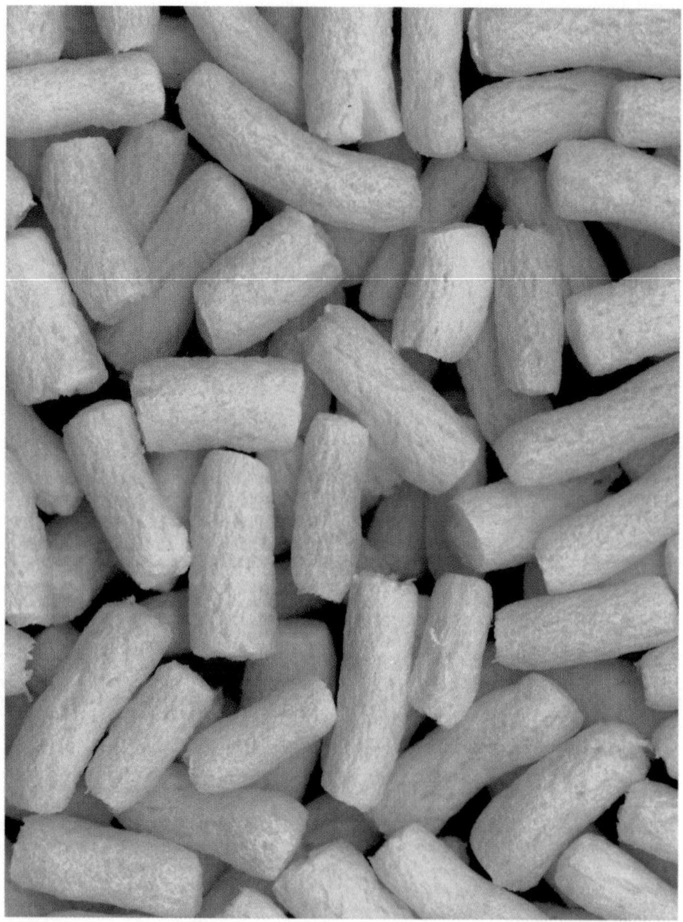

Eco-foam™

Corn Starch Foam: Advertisement

Nature Foam

Are you concerned about the effects of hazardous materials on the environment?

Now, there is an environmentally safe alternative to plastic foam. New Nature Foam is made from a renewable resource—corn—grown in America. This loose fill material cushions items during shipment and contains no harmful products that will endanger the environment.

If you don't want to reuse the foam, applicable disposal regulations may allow you to dispose of it in a compost pile, flush it down the toilet, wash it down the sink, or let the rain wash it away.

Why not use paper or popcorn for your packaging needs? Popcorn is messy, can turn rancid, has a higher weight per unit volume (it's more dense), and can attract rodents. Paper is heavy and requires extra time to pack into the box, increasing your costs of box packing and handling.

Switch now to the environmentally friendly Nature Foam. Although it does cost a little more per use than polystyrene foam, peace of mind makes it well worth it.

PROCEDURE

1. Set up the data table.
2. Read the background information and advertisements about the two kinds of foam. Look for evidence to help you make the decision. Record the evidence.
3. Decide which tests you and your partner will perform on the samples. You must do at least three tests.
4. Perform the tests on the polymer samples. Record the procedures you used and the results.
5. Check to see that you have done enough tests.

ANALYSIS

Begin the investigation write-up. Your report should include a data table to present your results. Be sure to answer the following questions in the conclusion:

1. Which polymer did you choose for the shipping department of Mike's Games?
2. What are the reasons for your choice?
3. Are there any disadvantages to your choice?
4. What do you see as the trade-offs in your decision?
5. What kind of evidence would help you make a better decision?

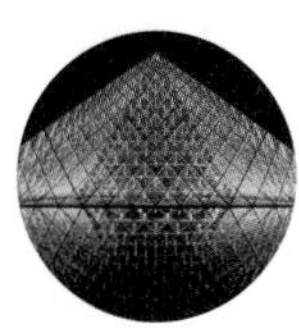

Solid Waste Decisions

Data about the volume and weight of solid wastes can provide useful information for making decisions about wastes. In this activity you will compare data about the garbage you generate with data about garbage collection nationally.

Compare your data with national data about garbage categories.

Landfill near Chicago, Illinois

PROCEDURE

1. Use the Student Sheet 11 to fill in the data about your personal garbage.
2. With your group, estimate the percentage of the total represented by each category. Record your group estimates on Student Sheet 11.

ANALYSIS

1. Compare your class data with the data from the national landfill graphs. How are they alike? How do they differ?
2. How would you explain the differences?
3. If waste is placed in a landfill, what is more important—the volume or the weight?
4. What are some of the major concerns when making a decision about what to do with the total waste that is accumulated in a community?

Composition of Municipal Solid Waste by Weight, 2000

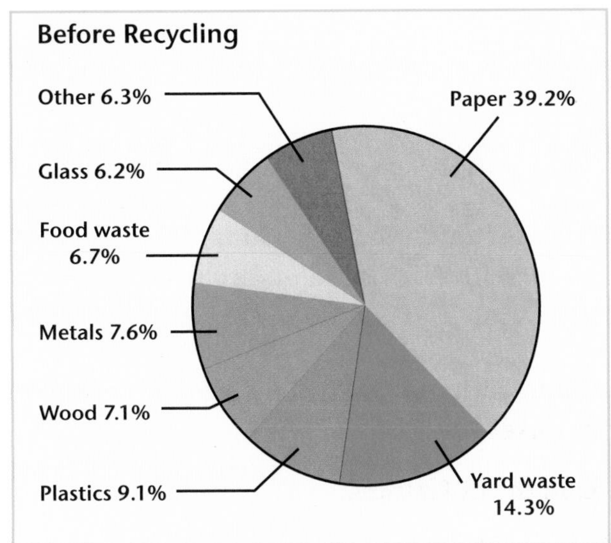

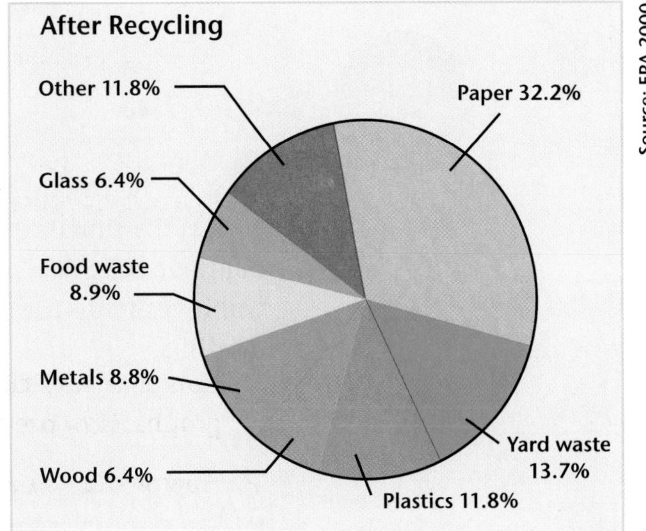

Composition of Municipal Solid Waste by Weight, 2000

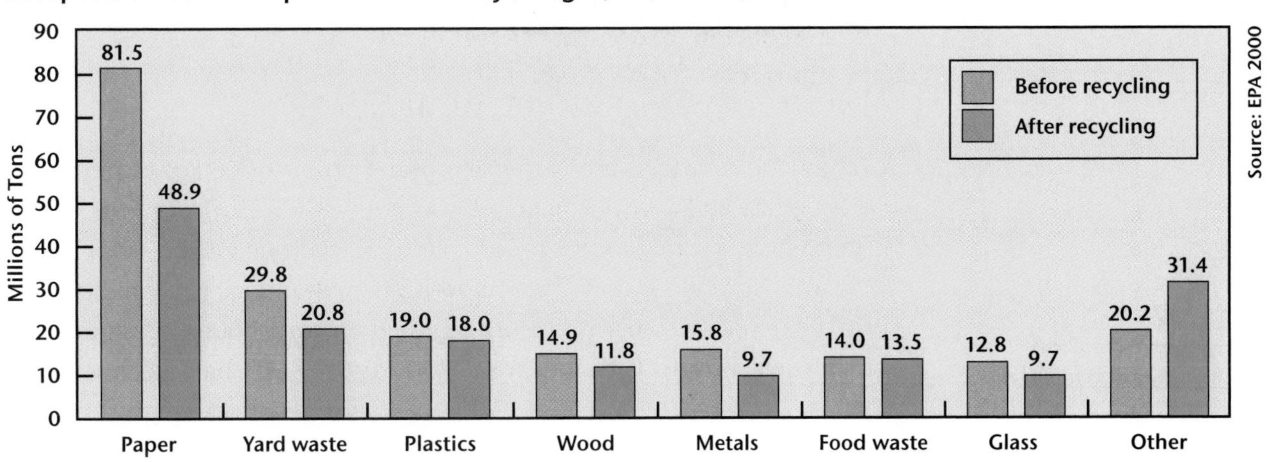

Landfills

Most of the municipal solid waste (MSW) produced in this country eventually finds its way to a landfill. But some landfills are operating near capacity; many others will close soon. In this activity, you will investigate the construction and operation of modern sanitary landfills.

CHALLENGE

Use the information in the reading, along with the investigations, to help you understand the advantages and disadvantages of sanitary landfills.

Part One: An Historical Look at Landfills

Today, the average American discards 4.4 pounds (2.0 kg) of municipal solid waste each day, or nearly 1,600 pounds (727 kg) per year. This is approximately twice the amount that the average citizen of Germany discards. As a nation, we discard over 221 million tons (154 million metric tons) yearly. Every day in the United States 63,000 garbage trucks find their way to the nation's 5,200 landfills. If the trucks were all arranged in a single line, the line would stretch over 373 miles (approximately 600 km) bumper to bumper. If current practices continue, our annual solid wastes will total nearly 250 million tons by the year 2010!

A solid waste transfer station

We live in a "throw-away" society. Whether we call it garbage, trash, refuse, or rubbish, it all boils down to the same thing: materials that someone no longer wants. As a society, we want all of our wastes to go away magically, but where is "away?" We want the products our technology creates, but not the wastes. This problem has resulted from the failure to connect the technological processes that we use to create products with the wastes that are created and have to be disposed of in some way. How did this begin?

Dumps and landfills have been the traditional answer to the problem of disposing of our wastes. In ancient Athens 2,500 years ago, it was decreed that wastes be transported at least one mile beyond the city gate. In colonial times, waste sites were located outside of villages, or the wastes were burned as fuel. These solutions were an improvement over earlier practices of dumping anywhere, but as the populations of cities increased and dumps grew larger during the Industrial Revolution, they soon created problems.

Some of the first attempts at constructing more modern landfills in the United States occurred in the Midwest in the early 1900s. Concern for public health was the main factor behind this change. By the 1930s, sanitary landfills were becoming more common. When the U.S. Army Corps of Engineers adopted sanitary landfills as the major disposal option for military facilities, many people were trained in their benefits and operation. Because of the odors and obvious pollution from the incinerators of the times, landfills came to be regarded as a highly desirable alternative. By 1960, about 1,400 American cities had sanitary landfills.

Today, modern landfills are carefully located, designed, and built to contain wastes and prevent their release to the environment.

QUESTION

How would you design a landfill that would address the pollution problems of the early dumps?

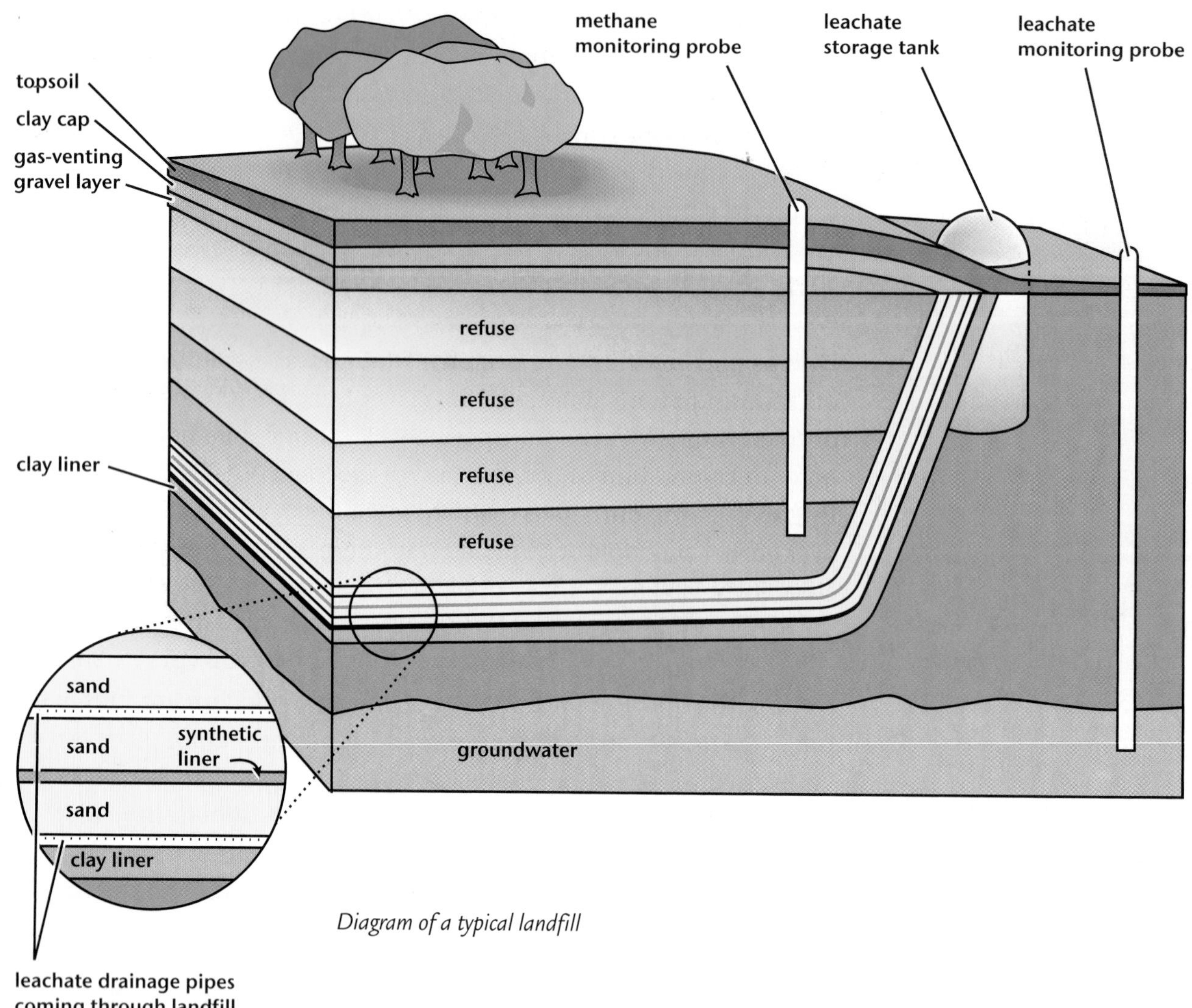

Diagram of a typical landfill

Part Two: What Happens in a Landfill?

Studies of the geology and water sources of an area are now used to determine where landfills can be located without contaminating ground and surface water. The landfills are usually lined with several feet of dense clay and then sealed with thick layers of plastic to prevent leaks. Each day, several inches of soil are spread over the landfill to prevent release of odors and to prevent rainwater from seeping in to form leaking materials called **leachates.** Leachates are liquids that seep from a landfill and contain dissolved substances of all kinds. They can be toxic or not, depending on the materials that have been discarded in the landfill.

As leachates collect in the bottom of the landfill, they are pumped out and then are collected and treated in a manner similar to sewage. The water is purified, and the remaining solids (a muddy mixture called sludge) are dumped in the ocean or a landfill, burned, or used as fertilizer. If the sludge is considered to be hazardous, it is sent to a hazardous waste disposal site.

Landfills also produce methane gas as a byproduct of the breakdown of organic substances, such as lawn and food wastes. The newest landfills contain a system of pipes to collect the methane gas. This gas is sometimes vented to the air or burned. In other cases, it is purified and used as a fuel.

Although reducing the amount of waste dumped into landfills is one of the goals of modern waste management policies, landfills will always be with us. Even recycling produces wastes that must be landfilled. Landfills are controversial because of where they are located; no one wants to live near one—the so-called NIMBY (Not in My Back Yard) syndrome.

QUESTIONS

1. What is the purpose of spreading soil over the landfill each day?
2. How are the leachates that collect in the bottom of the landfill treated?
3. Name the various ways of disposing of the sludge remaining from the treatment of leachates.
4. What is produced as a byproduct of the decay of organic wastes? How is it treated or disposed of?
5. What do you think should be considered when making decisions about where to locate landfills and how to manage the wastes? List the factors that should be considered and explain why each is important.

Part Three: Landfills and the Environment

Leachates from landfills can have serious consequences. They can contaminate drinking water sources or have other harmful effects on the environment. The potential impact of leachates on the environment is illustrated by what can happen when landfills are built on or near wetlands. Wetlands are low-lying areas that are damp or wet most of the year. Marshes and swamps are examples of wetlands commonly found in the United States. Grasses, weeds, and other non-woody plants are common in marshes. In swampy areas, bushes and trees are also present.

Until the 1970s, wetland areas were often considered to be undesirable. By the early 1970s, more than 30% of the wetlands in the United States had been drained or filled. Large areas of cities, such as Boston, New Orleans, Miami, and St. Louis, are built on filled wetlands. Landfills were also frequently located in wetland areas. The filled land could later be turned into desirable real estate. This was viewed as an added bonus of landfill construction.

Recently, our understanding of the ecological importance of wetlands has increased. We now know more about the impact of environmental changes in such areas. Wetlands act as a storage area for water, absorbing water during floods and acting as a reservoir of water in times of drought. In addition, wetlands serve as rich habitats for

Toxic substances can accumulate through food chains

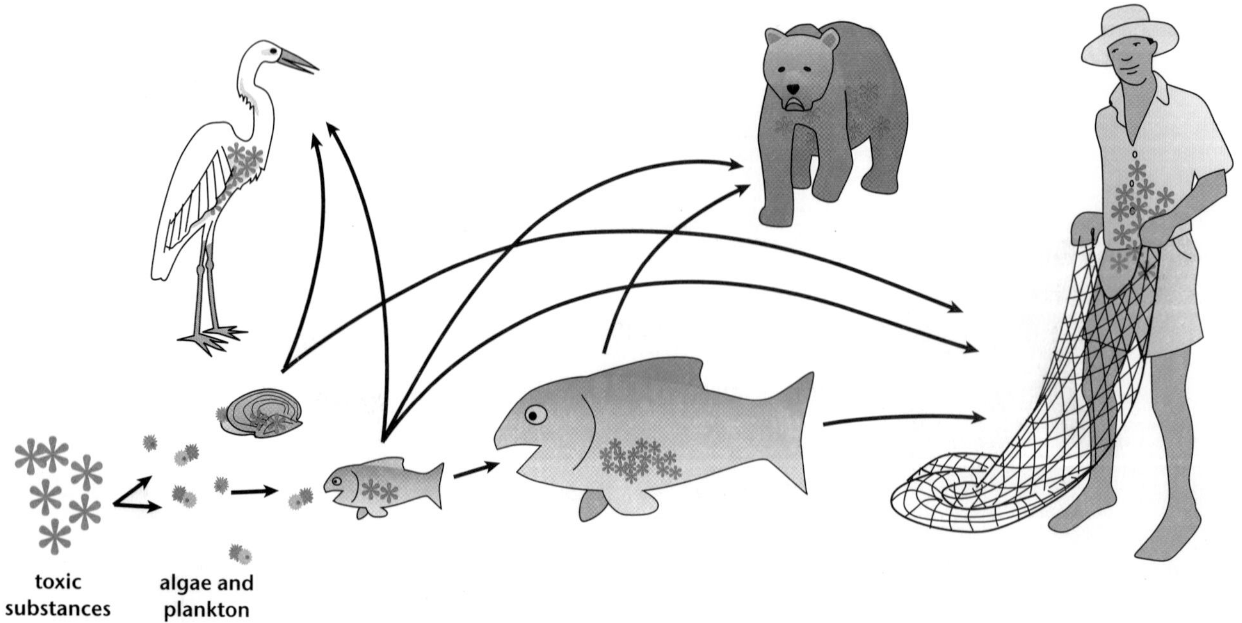

wildlife. Many species of animals, especially birds, obtain food and shelter and reproduce in wetland areas.

The leaching of toxic materials from landfills sited near wetlands, rivers, streams, and the ocean has contributed to the bad reputation landfills have in some areas. Once leachates enter a wetland or other aquatic areas, they are able to spread into the environment. The problem can be magnified if the toxic substances are concentrated by living organisms in the food chain. Concentration occurs when toxic substances are taken in by organisms at the base of the food chain, and organisms further up the chain eat large numbers of them. This is the case with some toxic heavy metals, like lead and mercury. Once they are taken in by algae and plankton, they cannot be broken down and excreted. They build up to higher concentrations in fish and invertebrates, like clams and crabs, that eat the smaller organisms. These are in turn eaten by birds and mammals, including humans. At that point, the levels of the heavy metals may have built up to harmful concentrations well above the threshold limits for humans and other organisms. The concentration of lead in fish can be 50 times greater than that in the water, and mercury can be concentrated up to 33,000 times!

Wetlands and other areas near surface water or aquifers are no longer considered appropriate places for landfills. Construction of landfills currently involves careful selection of a site as just one of many steps that must be taken to prevent release of leachates into the environment.

QUESTIONS

1. Until the 1970s what happened to many of the wetland areas in the U.S.?

1. Why are wetlands important?

3. What is the major problem resulting from landfills located near wetlands, rivers, and the ocean?

4. What greater problems result when toxic substances leach into aquatic areas?

5. What happens to toxic heavy metals when they are eaten by aquatic organisms?

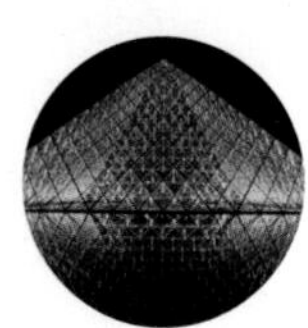

Making a Model of a Landfill

Some materials, when they are thrown away carelessly, can leak into the surrounding soil and eventually get into the groundwater. These leaking materials are called leachates. A variety of precautions are taken to prevent materials from leaking out of landfills.

Examine leachates from both a simulated unregulated dump and a regulated landfill in order to understand issues related to the safety of landfills.

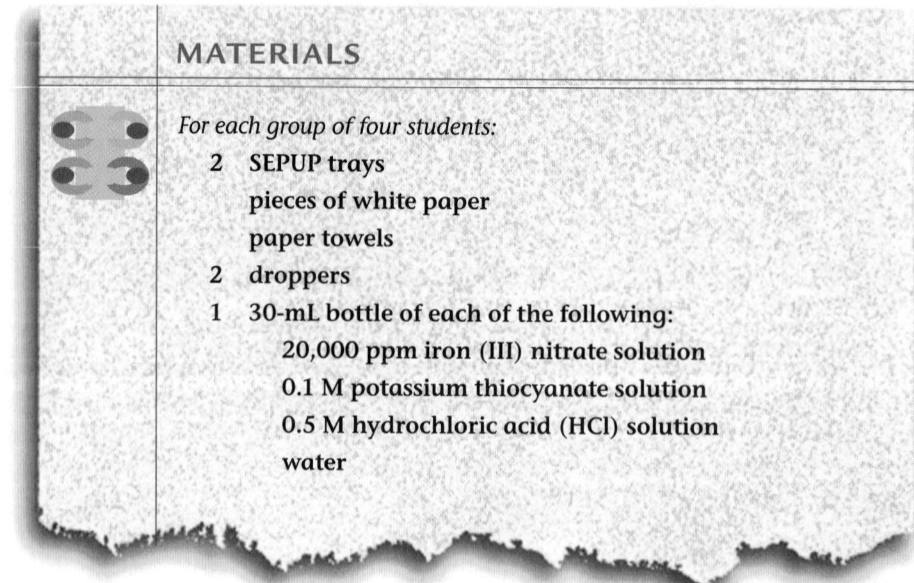

MATERIALS

For each group of four students:

2 SEPUP trays
pieces of white paper
paper towels
2 droppers
1 30-mL bottle of each of the following:
- 20,000 ppm iron (III) nitrate solution
- 0.1 M potassium thiocyanate solution
- 0.5 M hydrochloric acid (HCl) solution
- water

PROCEDURE

Your teacher will set up a demonstration of two leachate tubes to simulate an unregulated dump and a sanitary landfill. Iron (III) nitrate is used to simulate a toxic heavy metal. Student pairs will test the leachate samples prepared by your teacher. In order to determine the concentration of heavy metal in the leachates, you will need to set up a serial dilution of iron (III) nitrate to serve as a standard. You will determine the iron (III) nitrate concentration in the leachates by comparing them to the standards.

SAFETY NOTE: Do not taste or touch the solutions. Wash any affected area with water for 2–3 minutes. Wear safety eyewear.

1. Prepare a serial dilution of the 20,000 ppm iron (III) nitrate in Cups 1–5 of your SEPUP tray. Start with 10 drops of 20,000 ppm iron (III) nitrate in Cup 1. When you have completed the dilution procedure, record the concentration and color of each dilution in Cups 1–5.

2. Obtain 10 drops of each of the leachates prepared by your teacher. Add these to Cups 6 and 7 of the SEPUP tray. Record the color of the leachates.

3. Test for iron by adding one drop of potassium thiocyanate to each of Cups 1–7. Record the color of the solution in each cup. Then add one drop of hydrochloric acid to each cup and record your observations.

4. Determine the concentration of iron (III) nitrate in the leachates by comparing them to the standards in Cups 1–5.

5. Set up a data table to organize and present your results. For each solution, you should record the color, both before and after testing, and the ppm of iron detected.

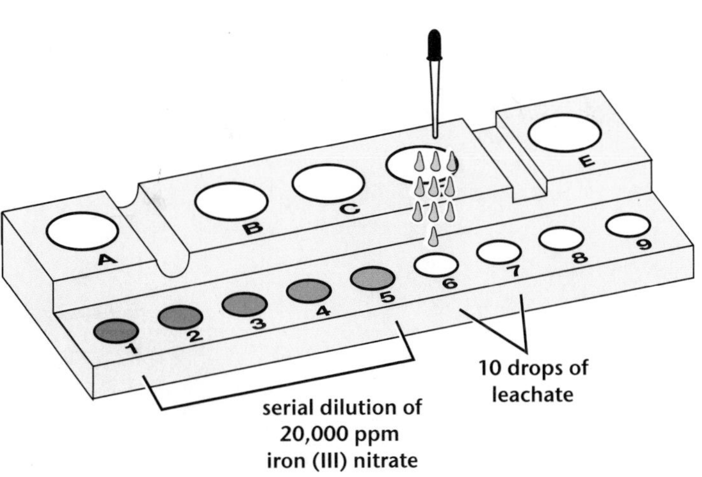

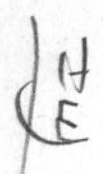

ANALYSIS

1. Observe the leachate tubes prepared by your teacher. What does the gravel/water layer represent?
2. What does the clay layer represent?
3. Explain what you can conclude about leachates from the simulated dump and landfill activity.
4. What risks, if any, are there in locating a dump on or near an aquifer that is used as a source of drinking or agricultural water?
5. What risks, if any, are there in locating a sanitary landfill on or near an aquifer used as a source of drinking or agricultural water?
6. Is it possible to operate a landfill that is completely risk free? Explain.
7. What materials would you dispose of in a sanitary landfill? Explain.
8. What personal action can you take to help make sure that wastes are disposed of properly—for example, that possibly hazardous substances are not put in sanitary landfills where they may endanger public safety?

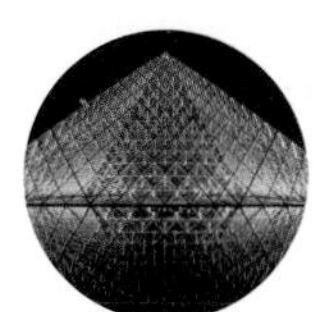

Incinerating Heavy Metals

You will observe a simulation of the incineration of a hazardous waste that includes a toxic heavy metal. Then you will examine and test both the gaseous and solid products of the incineration.

CHALLENGE

As you investigate the products of incineration think about the advantages and disadvantages of hazardous incineration. Also think about what can be done to minimize the risks of operating a hazardous waste incinerator.

A hazardous waste incinerator

PROCEDURE

Part One: Up in Smoke

1. Your teacher will burn a piece of paper that contains simulated heavy metals and will collect the ashes. Observe the demonstration, record your observations, and answer Questions 1–3 on the next page in your science notebook.

2. Your teacher will burn a second piece of paper that contains simulated heavy metals and will collect both the ashes and the smoke.

3. In your science notebook record the results of your teacher's demonstration to test the smoke with bromothymol blue (BTB) indicator. Answer Question 4 and discuss your answer with your classmates.

4. In your science notebook record the results of your teacher's demonstration to test the smoke with the potassium thiocyanate. Answer Question 5.

5. After viewing the videotape, "Incinerating Hazardous Wastes," answer Questions 6–8.

Controlled high-temperature combustion—such as occurs in a modern power plant or incinerator—is much more efficient than burning wood in a campfire.

ANALYSIS

1. What are the two main products that come from the incineration of the heavy metal paper?

2. How effective was incineration in reducing the volume of the paper? What is your evidence?

3. How would you prove that the ash left behind had changed in weight? Do you think it weighs more, less, or the same as the unburned paper?

4. Why do you think the water that came from the bag used to collect the smoke behaved as it did with bromothymol blue (BTB) indicator?

5. Why did the water that came from the bag used to collect the smoke behave as it did with the potassium thiocyanate and hydrochloric acid?

6. What conclusions can you draw about the contents of the smoke based upon what you have seen in class?

7. What problems do the gases formed in incineration pose to the environment? What is a possible solution?

8. What problem does the ash formed in incineration pose to the environment? What is a possible solution?

PROCEDURE

Part Two: Investigating Residues

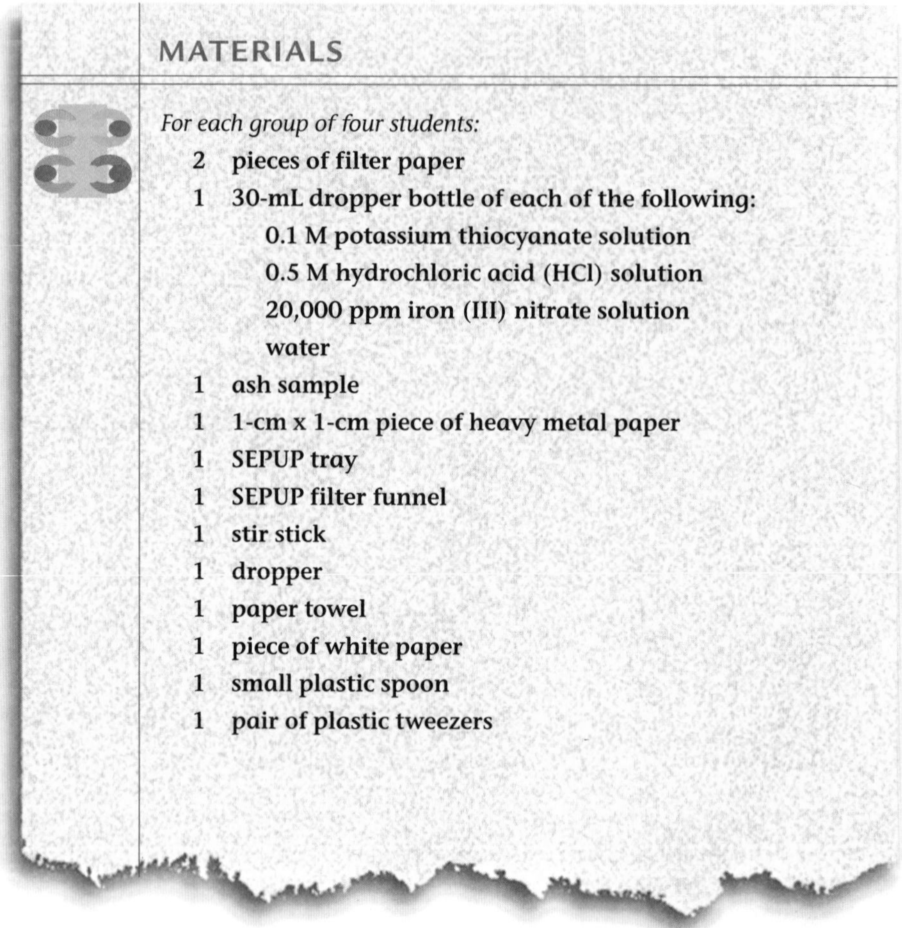

MATERIALS

For each group of four students:

2 pieces of filter paper
1 30-mL dropper bottle of each of the following:
  0.1 M potassium thiocyanate solution
  0.5 M hydrochloric acid (HCl) solution
  20,000 ppm iron (III) nitrate solution
  water
1 ash sample
1 1-cm x 1-cm piece of heavy metal paper
1 SEPUP tray
1 SEPUP filter funnel
1 stir stick
1 dropper
1 paper towel
1 piece of white paper
1 small plastic spoon
1 pair of plastic tweezers

SAFETY NOTE: Follow your school policy regarding the use of safety eyewear. Do not taste or touch the solutions. Wash any affected area with water for 2–3 minutes.

Read through the entire experiment before you begin and prepare a data table to record your observations.

1. Fill large Cup A half full of water. This cup will be used for cleaning the stir stick and dropper.
2. Place your SEPUP tray on a piece of white paper. Use the tweezers to put a 1-cm x 1-cm piece of unburned heavy metal paper in small Cup 1. Add 10 drops of water. Stir slowly for one minute.
3. Add one drop each of potassium thiocyanate and hydrochloric acid solutions to Cup 1. Stir the mixture and record your results in your data table. Clean your stir stick.
4. Add a half spoon of ashes to small Cups 3 and 5. Break them into small pieces with your stir stick. Add 20 drops of water to Cup 3. Stir.
5. Set up the SEPUP filter funnel over large Cups B and C. Fold two pieces of filter paper as directed by your teacher and shown in the drawing below, and place these in the funnels. Wet each thoroughly with 20 drops of water.

Step 1: Fold filter paper in half.

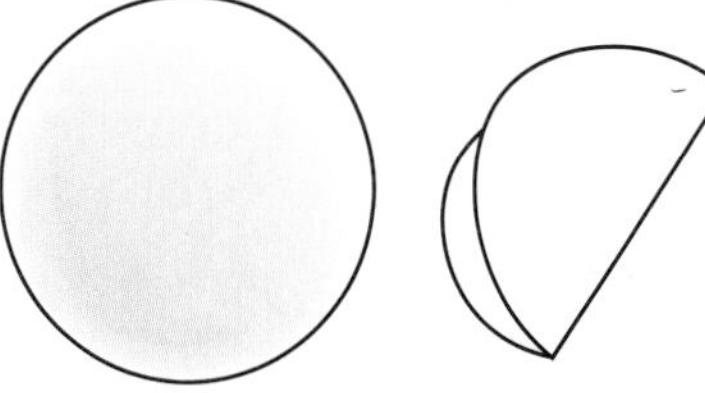

Step 2: Fold filter paper in half again.

Step 3: Open with three thicknesses of paper on one side of cone and one thickness on the other.

Step 4: Place in funnel and wet paper with water to hold it in place.

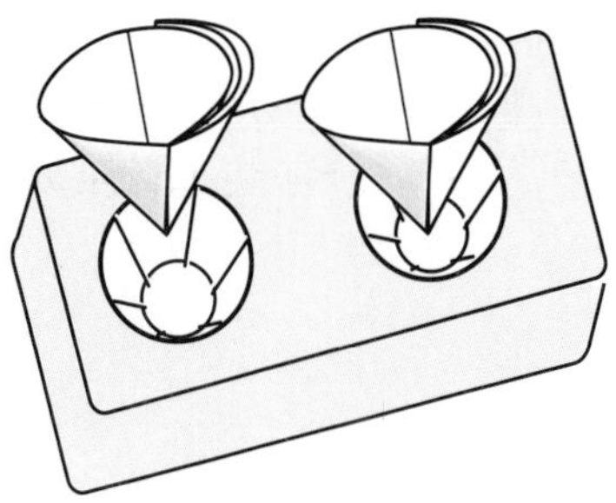

6. With the dropper, remove as much liquid as possible from small Cup 3 and place it in the filter funnel over large Cup B. Clean the dropper. Allow the material to filter while you do the next part of the activity.

7. Add 20 drops of hydrochloric acid solution to small Cup 5. Stir. With the dropper, transfer all of the liquid in small Cup 5 to the filter funnel over Cup C. Allow the material to filter for 2 to 3 minutes.

8. Remove the SEPUP filter funnels to a paper towel. Discard the filter papers in the trash and clean and dry the filter funnels.

9. Record the color of the filtrates in large Cups B and C.

10. Add two drops of potassium thiocyanate solution to large Cups B and C. Add 1 drop of hydrochloric acid to large Cup B. Stir Cups B and C. Record the colors in your data table.

11. Use your data table from Activity 12 to find the approximate heavy metal concentration in each cup. Record this in your new data table.

12. Clean up as directed by your teacher and then answer the Analysis questions.

ANALYSIS

1. How did the hydrochloric acid affect the heavy metal paper ash?
2. How does the concentration of heavy metal found in the ash and water filtrate compare to the concentration of heavy metal found in the ash and hydrochloric acid filtrate? How do they both compare to the unburned paper heavy metal concentration?
3. If incinerator ash contains toxic substances, such as heavy metals, it is commonly disposed of in special hazardous waste landfills. What are some problems these special landfills must solve in order to avoid the risk of contaminated groundwater? What can be done to solve these problems?
4. Batteries, some paints, and some pigment colors used in plastics contain heavy metals. What disposal policy would you adopt for your town if these materials were sent to a community sanitary landfill? What if they were sent to a hazardous waste incinerator?

GOING FURTHER

Find out what methods, if any, are available in your area for the treatment and disposal of hazardous waste ash from heavy metals.

14 Recycling Materials

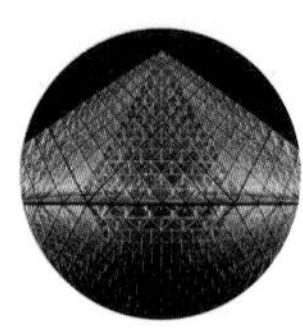

Investigating a New Plastic

Recycling and reusing materials is an important component of meeting our goal to reduce the amount of waste we throw away. This investigation looks at a new plastic that has some very unusual properties that enable it to be reused. It is used as a model for the recycling of materials generally.

Plastic containers to be recycled

CHALLENGE

Observe the new plastic's unusual properties by testing it in an acid and a base. Based on your observations, determine whether or not there might be environmental benefits associated with the use of this type of plastic.

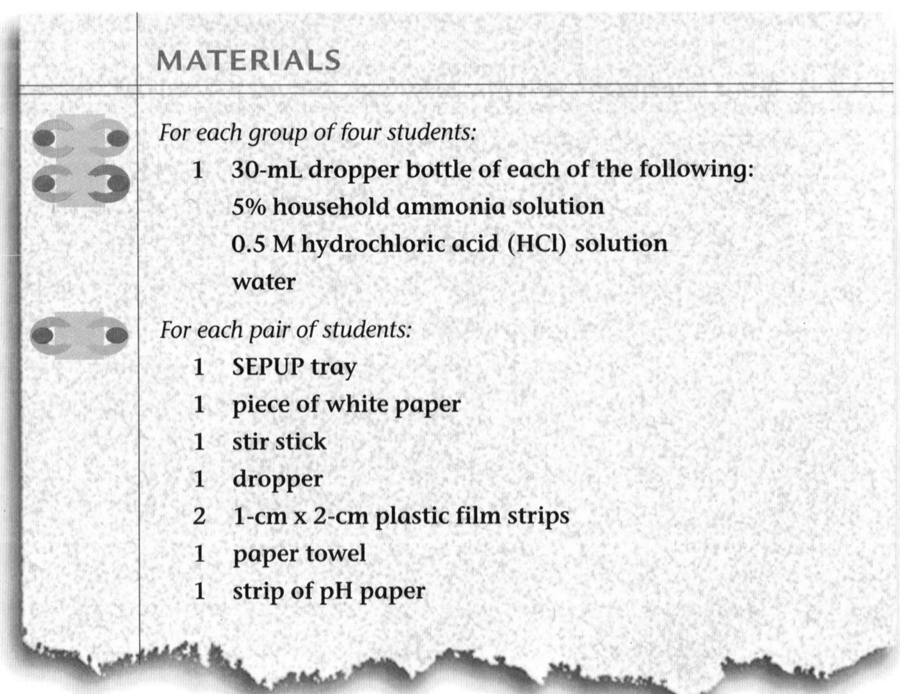

MATERIALS

For each group of four students:

1 **30-mL dropper bottle of each of the following:**
5% household ammonia solution
0.5 M hydrochloric acid (HCl) solution
water

For each pair of students:

1 **SEPUP tray**
1 **piece of white paper**
1 **stir stick**
1 **dropper**
2 **1-cm x 2-cm plastic film strips**
1 **paper towel**
1 **strip of pH paper**

PROCEDURE

SAFETY NOTE: Do not taste or touch the solutions. Follow your school policy regarding the use of safety eyewear. Wash your hands after completing the activity.

Read through the entire procedure before you begin and make a data table to record your observations.

1. Describe the plastic film (color, density, flexibility, the way light is reflected from it). Record your observations.
2. Place one of the pieces in large Cup A of your SEPUP tray. Add 50 drops of water. Stir with the flat end of the stick. Put a drop of water from Cup A on a piece of pH paper to check the pH of the liquid. Record your observations of the plastic film and the color of the pH strip.
3. Place the other piece of plastic in large Cup B of your SEPUP tray. Add 50 drops of household ammonia to large Cup B. Stir with the flat end of your stir stick until no further change takes place.
4. Add a drop of the liquid in Cup B to a piece of pH paper to check the pH. Record your observations of the plastic-ammonia mixture and the color of the pH paper.
5. Use the dropper to transfer the liquid contents of large Cup B to large Cup C.
6. While stirring, add 40 drops of hydrochloric acid solution to Cup C. Place a drop of the liquid in Cup C on a piece of pH paper to check the pH. If its color is green or blue, add 5–10 more drops of hydrochloric acid and retest the solution with another piece of pH paper. Repeat, as necessary, until the pH paper color is orange or red. You will know when you have been successful. You will see the mixture change from a milky white color to a lightly cloudy to a clear color. Record your observations.
7. Remove all of the reclaimed plastic, wash it and place it in large Cup D. Some of the plastic may stick to the stir stick.
8. Add 50 drops of water to Cup D. Touch a small piece of pH paper to the solution in Cup D. Record your observations.

9. Use the chart with the pH paper to estimate the pH of each of the solutions. Record your observations.

10. Remove the plastic from Cup D and place it between layers of the paper towel. Press down to squeeze out the water. Examine your recycled plastic. Try pressing it against a penny or shaping it with your hands. Record its appearance and properties in your data table.

11. Remove the plastic from Cup A to the paper towel. Compare the properties of this plastic to your reclaimed/recycled plastic. Record the differences in appearance and properties. Answer Questions 1–4.

12. Allow your recycled plastic to dry overnight and then re-examine its properties. Record its appearance and properties. Answer Questions 5–7.

ANALYSIS

1. What affect does water alone have on the plastic? Why do we put the plastic in the water?

2. What happens to the plastic when ammonia is added?

3. What happens to the solution in Cup C when hydrochloric acid is added to it?

4. Write a sentence or two describing the relationship between pH and the properties of the plastic.

5. Compare the properties of the recycled plastic to those of the original. Did the recycling process affect the plastic in any way?

6. Could your recycled plastic be used again as a bag? What other uses might the recycled plastic have?

7. Do you think the solutions in Cups B, C, and D can be added directly to the city waste water (sewer) system of your community? Why or why not?

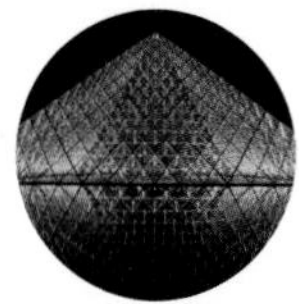

Plastics Recycling Today

Effective recycling of plastics requires some special approaches by the manufacturer, the consumer, and the recycler. Through watching and discussing a video and the sorting of plastic containers you bring into class, you will become more knowledgeable about plastics recycling.

CHALLENGE

After viewing the videotape, "Plastics Recycling Today: A Growing Resource," answer the following questions in your science notebook. Think about what you want to discuss regarding plastics recycling.

QUESTIONS

1. What are the advantages and disadvantages of making product containers from plastic compared to using other materials, such as glass or aluminum?
2. Write a sentence or two describing each phase of the plastics recycling process: collecting, sorting, processing, and remanufacturing the recycled plastic into useful products.
3. Is there a recycling center near your home that recycles plastic? Do you recycle plastics? Describe what you do to recycle plastics, or why you don't recycle them.
4. Is recycling plastics a good idea? Give reasons to support your position on this issue.
5. Using the codes and descriptions in the chart on the following page, sort the containers and/or records of containers you brought from home.

Recycling Plastics			
Code	**Name**	**Common Uses**	**Examples of Recycled Products**
1	PETE—polyethylene terephthalate (PET)	soft drink bottles, peanut butter jars, salad dressing bottles	Liquid soap bottles, fiberfill for winter coats, paint brushes, soft drink bottles, film, egg cartons, skis, carpets, boats
2	HDPE—high density polyethylene	milk, juice, and bleach bottles; grocery bags; toys; liquid detergent bottles	flowerpots, drain pipes, toys
3	V—vinyl, polyvinyl chloride	shampoo bottles, clear food packaging, pipes	pipes, floor mats, hoses, mud flaps
4	LDPE—low density polyethylene	plastic films, bread bags, frozen food bags, grocery bags	garbage can liners, grocery bags, multi-purpose bags
5	PP—polypropylene	ketchup bottles, yogurt cups, margarine tubs, medicine bottles	paint buckets, ice scrapers, fast-food trays, automobile battery parts
6	PS—polystrene	videocassette cases, compact disc jackets, grocery store meat trays, coffee cups, prescription bottles, utensils	license plate holders, flowerpots, hanging files, food service trays, trash cans
7	Others, including mixed polymers	packaging	plastic lumber

Are We Recycling Enough?

Recycling prevents useful materials from being burned in incinerators or buried in landfills. It can save energy and natural resources as well as saving landfill space. According to data from the EPA (2000), national recycling efforts divert about 27% (60 million tons) of municipal solid waste from landfills.

CHALLENGE

Recycling delays the day when a material's useful life finally ends. Why don't we do more of it? Think about ways to encourage people to do more recycling.

Materials are separated for recycling

ENCOURAGING RECYCLING

Recycling prevents useful materials from being burned in incinerators or buried in landfills. It can save energy and natural resources, provide useful materials from discards, and save landfill space.

Municipal solid waste can be thought of as a mixture of potentially valuable materials and fuels. Glass, plastics, paper, and metals are excellent candidates for recycling. Yard wastes can be composted and added back to the soil or incinerated for energy. Yet large amounts of recyclables are still ending up in incinerators and landfills, where they account for a significant percentage of the total waste. For example, even though 50% of all aluminum cans are recycled, we still discard enough aluminum to rebuild our entire domestic air fleet each year. Why is this?

The answer depends on a number of factors, including the amount of effort recycling requires, the costs of recycling, and sometimes the lack of a market for the recycled products.

Communities must examine the costs and benefits of recycling, considering questions such as: What markets exist for this material? How do the costs of recycling compare to the costs of incinerating or putting the wastes in landfills? For many cities, paying to recycle paper may be less expensive in the long run than putting it in landfills. Local, state, and national governments are increasingly aware of this, and some have begun to require the purchase of recycled paper. For example, in California, newspapers must contain at least 25% recycled newsprint.

Many cities and states are taking innovative approaches. Because most yard wastes can be composted, some states no longer allow it to be placed in landfills. In other places landfills are being fitted with giant shredders to grind yard wastes into compost that can be distributed to citizens or sold as fuel. In many communities nationwide, citizens are charged directly for the amount of trash they generate. For example, they pay one rate for their first can. Additional cans cost extra. The result of this policy is an awareness of the economic costs of disposal. The policy has also helped encourage residents to recycle, compost, and reuse.

QUESTION

Historically, cars have been dumped in junk yards when their useful life is over. Based on everything you have learned about waste management in the last few activities, suggest a complete plan for the disposal of a car. Be sure to consider the car's metal body, plastic bumpers, lead-acid battery, interior, and other components.

Discarded cars in a junkyard

15 The Aluminum-Copper Reaction

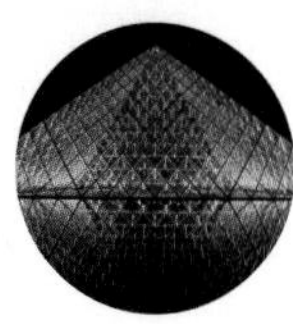

Exploring Chemical Reactions

When two or more materials are mixed, they sometimes change to form new materials. This is called a chemical reaction. You will explore chemical interactions by examining the reaction between aluminum and a solution of copper chloride. This is an example of a single replacement reaction that can be used to recover copper for reuse.

The silver from chemicals used to develop film is reclaimed and reused because it is too toxic (and too valuable) to dispose of in the sewer system.

CHALLENGE

Find as much evidence of chemical interaction as you can in this reaction. How do the properties of the reactants and products of a chemical reaction compare to each other?

MATERIALS

For each group of four students:

- 1 SEPUP tray
- 1 graduated container
- 1 10-cm x 10-cm piece of aluminum foil
- 2 9-oz. clear plastic cups
- 1 plastic spoon
- 1 wooden or metal spoon
- 1 dropper
- paper towels
- 1 180-mL bottle of 50,000 ppm copper chloride solution
- 1 30-mL dropper bottle of each of the following:
  - 50,000 ppm copper chloride solution
  - 5% household ammonia solution
  - water

PROCEDURE

SAFETY NOTE: Wear protective eyewear. Some people may have an allergic reaction to the copper chloride solution, resulting in itching and redness in the affected area for a short time. Wash any exposed area with water for 2–3 minutes.

1. Before you proceed, record some properties of the aluminum foil and the 50,000 ppm copper chloride solution in your science notebook.

2. Carefully fold the aluminum foil in half, and then in half again. Place this in the bottom of one of the plastic cups.

3. Fill a second plastic cup 1/3 full of water. Place the plastic cup with the foil inside this cup.

4. Use the graduated container to measure 25 mL of copper chloride solution from the 180-mL bottle. Pour it onto the aluminum foil in the clear plastic cup.

5. Use the spoon to push the corners of the aluminum down so that the foil is completely covered. As the reaction proceeds, record as many observations as you can. Try not to include any interpretations.

6. While the reaction proceeds, set up a 1 to 10 serial dilution (as you did with the iron III nitrate in Activity 12) of the copper chloride solution. Start by putting 10 drops of 50,000 ppm copper chloride from the 30-mL dropper bottle in Cup 1 of a SEPUP tray. Continue the dilution through Cup 5. After the aluminum-copper chloride reaction seems to be finished (about 5–10 minutes), use a dropper to add 10 drops of the liquid from the plastic cup containing the aluminum reaction to Cup 7 of the SEPUP tray.

7. Add 5 drops of household ammonia to Cups 1–5 and Cup 7. What do you observe? Record your results in your science notebook using a data table like the one on the next page.

Testing for Copper Concentrations

	Color of Solution		
Cup	After Water Dilution	After Adding Ammonia	Concentration (in ppm)
1			50,000
2			
3			
4			
5			
6			
7			approximately ________

8. Use your dropper to carefully remove the rest of the liquid from the plastic cup containing the aluminum reaction. Put it into a container provided by your teacher.

9. Using the plastic spoon, scoop up as much of the solid material from the clear plastic cup as possible onto a folded paper towel. Remove as much liquid as you can. Then, use the bottom of a wooden or metal spoon to press the solid material together. Observe and record what you see.

ANALYSIS

1. Copper compounds usually have a blue or green color. Turquoise is a compound of copper. How effective does the aluminum appear to be in removing the copper from the copper chloride solution? What is your evidence?

2. Describe the solid material. What do you think it is? Why? What is your evidence?

3. How would you go about proving or disproving that the gas generated in the aluminum-copper chloride reaction was water vapor (steam)?

4. What evidence did you observe that would help you prove a chemical reaction took place? Give three examples.

It's Elementary and Shocking!

What are the smallest things you can think of? In this reading you will discover how the idea of an atom was developed.

CHALLENGE

As you read "A Small, Powerful Idea," try to imagine how people 2,000 years ago developed the idea of atoms.

A SMALL, POWERFUL IDEA

The idea of atoms arose long ago in the minds of philosophers from countries as far apart as India and Greece. Imagine the following conversation in Greece around 420 B.C.

"Have you heard the latest? A group of our fellow Greeks say that everything we see is made of small particles called atoms! They say these atoms are so small that we cannot see them with our eyes. They claim that everything is made of empty space filled with atoms in constant motion, like tiny dancing particles of dust in a sunbeam! They believe that water and iron are different because they are made of different kinds of atoms. Could this be true, my friend? How can reason tell us about something we cannot see?"

The Greek Atomists were a group of scholars who proposed that there were simple particles called atoms. They believed that what we call matter could be broken down into smaller and smaller bits until we arrived at a bit that was no longer divisible or able to be cut. They called this an atom—from the Greek word, *atomos,* which means uncuttable. The Atomists thought the bit-sized particle still retained all the physical and chemical properties of larger chunks of matter, such as a chunk of gold. They believed that the world was made of different kinds of atoms. By combining and recombining with each other, these different atoms produced material things and all the changes we see taking place in them. You may have never thought of this before, but when you ask such questions as: "Is this ring really gold?," "What's in my pizza?," "Why do things corrode?," and, "Why do I change and grow older?," you are really asking the same question as the early Greeks, "What is the world made of and why does it change?"

Ideas Lost and Found

For over 2,000 years the concept of atoms was not widely accepted. During those years, early chemistry developed from the work of alchemists, who believed that all metals were essentially the same. The alchemists devoted their lives to searching for ways to turn common metals such as lead into valuable silver and gold. Alchemists in Egypt, Greece, Turkey, and other parts of the Middle East laid the foundation for work that continued throughout Europe and Asia into the 18th century. Although these early scientists never did find a way to turn lead or iron into gold, some of them made important contributions to the study of science. For example, early Islamic scientists are credited with introducing careful record keeping and controlled experiments.

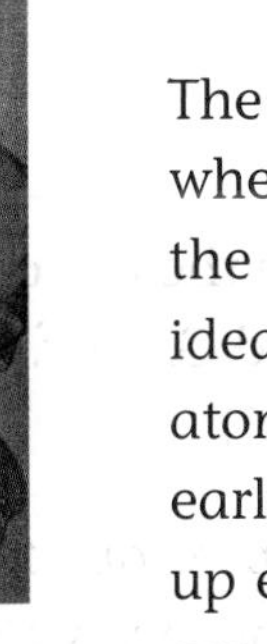
John Dalton (1766–1844)

The ideas of the Atomists were revived in the 13th and 14th centuries when Islamic and Jewish scholars in Spain rediscovered and translated the books of the early Greeks. The theories of the Atomists and the ideas and work of other scientists laid the foundation for modern atomic theory, which began to be developed by John Dalton in the early 1800s. Today, we believe that the individual particles that make up everything—including pizzas, stars, watches, whales, and water—are composed of atoms.

It's Elemental

Materials that are composed entirely of only one kind of atom are called elements. There are now more than 110 different elements known to exist. Some of these elements only exist when produced by scientists in laboratories. Only nuclear reactions can produce new elements or change one element into another. The alchemists were unable to turn lead into gold because lead and gold are different elements and, therefore, composed of different atoms. Although atoms cannot be changed into other atoms by normal chemical reactions, they can and do combine with each other to form molecules. In fact, elements are rarely found in their pure state in nature (gold, silver, platinum, copper, sulfur, and carbon are the only ones). Atoms are usually combined with atoms of a different kind. Materials that contain one or more different kinds of atoms are called compounds. Water, for example, is a compound made up of two hydrogen atoms and one oxygen atom. A single particle of the compound water is called a water molecule. There are so many compounds, and new ones being made all the time, that it would take many large books with tiny type just to list them!

Your body is made from many different combinations of a very few common elements. These are mainly hydrogen, oxygen, carbon, nitrogen, phosphorous, and sulfur. Other elements, such as calcium, magnesium, iron, sodium, potassium, and iodine, play an important role in your body's chemical energy plant, but they are used in smaller amounts.

Now you can answer the ancient question that the Greeks posed over 2,400 years ago. The world is made of atoms that are in constant motion. It changes because these atoms combine with each other to make compounds. In turn, the atoms in compounds can break apart and recombine to form new compounds that may last for less than a second or for many thousands of years before they take part in new reactions.

Positive and Negative

As scientists probed the atom more closely, they discovered that it was composed of even smaller particles that had electrical charges. Just as the poles of a magnet have north and south poles, atoms have positive and negative charges. The negatively-charged particles are called electrons. The positively-charged ones are called protons.

Simplified diagram of the structure of an atom

It was determined that electrons are found near the outside of an atom, and the protons are found near its center. This would help explain an experience you have had many times. When objects are rubbed (like your shoes on the carpet on a dry day), they are able to transfer electrons easily because the electrons are found near the outside of atoms. When these electrons jump, you may get a shock or see a spark fly. This also happens when you pull a wool blanket away from a cotton sheet on a dry day.

So now we know that atoms are made of smaller particles called electrons and protons. (A third particle that is found with protons in the center, or nucleus, of an atom is a neutron. It is neutral—neither positive nor negative.) Normally, atoms (elements) have the same number of positive and negative charges; if they did not, we would receive a shock every time we touched anything! But elements, as we said before, are rarely found in their pure form in nature. Most elements combine with each other to form compounds in a chemical change.

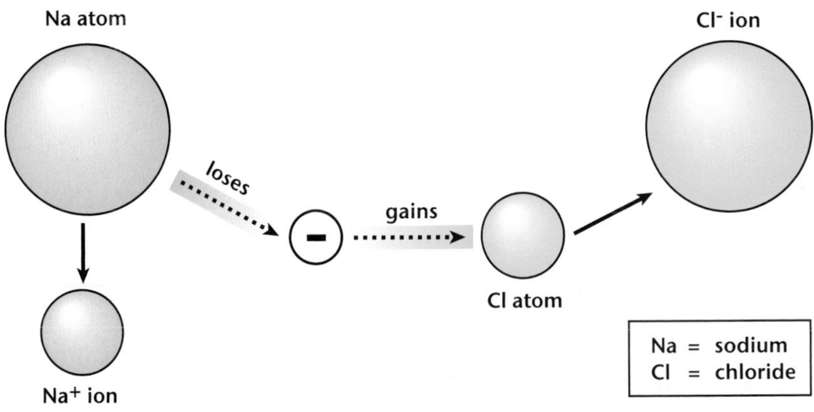

Formation of ions

One way this happens is by atoms losing or gaining electrons from each other. Common table salt, for example, is made up of sodium and chlorine atoms, but they have changed their elemental properties enormously. Sodium is a soft metal that explodes when it is immersed in water. Chlorine is a toxic, greenish-yellow gas. When these two elements come in contact with one another, each sodium atom loses an electron to a chlorine atom.

As you can see in the diagram, sodium, which was neutral before, has one less electron. Before, it had the same number of electrons (− charges) and the same number of protons (+ charges). Now there is one less electron than protons. We say that the sodium atom has a net positive charge, and we no longer call it an atom but an ion. An ion is a charged atom (or group of atoms). We call this ion a sodium ion.

What happened to the chlorine? Remember, it has gained the electron that sodium lost. What will be the net charge on chlorine when it gains an extra electron? It will have a net negative charge, and we call it a chloride ion. The positively-charged sodium ion and the negatively-charged chloride ion now attract each other to form the compound NaCl, sodium chloride, or ordinary table salt.

This process of materials losing and gaining electrons takes place in chemical reactions constantly. In your experiment with aluminum foil and copper chloride, aluminum atoms lost electrons to become aluminum ions. Copper ions gained electrons to become copper atoms—the reddish-brown solid in your cup.

QUESTIONS

1. Explain how an element can become an ion.
2. What is the difference between an element and a compound?

Investigating Metal Wastes

In Activity 15, you found that the metal aluminum removed copper from a toxic copper solution. Similar methods can be used to reclaim heavy metal wastes. Reclaiming copper serves a dual purpose—it reduces the toxicity of the waste stream and conserves a valuable metal. However, reclaiming the copper involves using another metal, which involves costs as well as the risks associated with disposing of the metal.

CHALLENGE

Investigate the effectiveness of three metals in reclaiming copper waste from a solution. Using the results of your investigation and additional information about the metals, you will choose a metal to use for reclaiming copper waste.

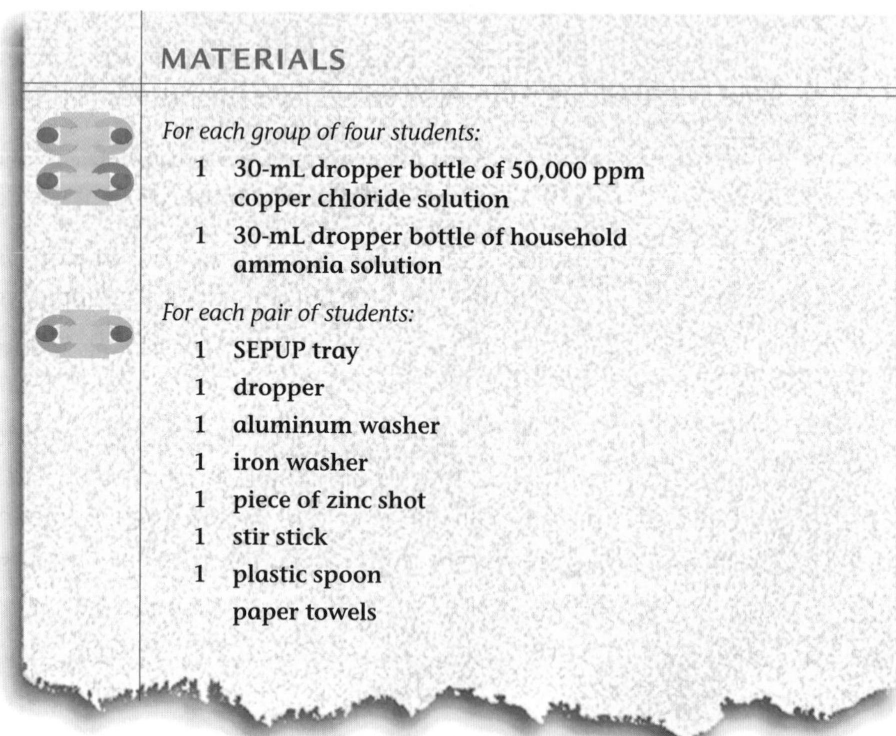

MATERIALS

For each group of four students:

1 **30-mL dropper bottle of 50,000 ppm copper chloride solution**
1 **30-mL dropper bottle of household ammonia solution**

For each pair of students:

1 **SEPUP tray**
1 **dropper**
1 **aluminum washer**
1 **iron washer**
1 **piece of zinc shot**
1 **stir stick**
1 **plastic spoon**
paper towels

PROCEDURE

SAFETY NOTE: Wear protective eyewear. Copper chloride solution is toxic and corrosive. Avoid contact with skin and eyes. Wash any exposed area with water for 2–3 minutes. Some people may have an allergic reaction to the copper chloride solution, resulting in itching and redness in the affected area for a short time.

1. Place the aluminum washer in Cup 1. Put the zinc shot in Cup 2. Put the iron washer in Cup 3.
2. Add 20 drops of 50,000 ppm copper chloride solution to each of Cups 1–4. Cup 4 will be used for comparison purposes.
3. Stir the contents of each cup every minute or so. Clean the stir stick between each cup.
4. Observe what happens to each metal for 5–10 minutes. Record your observations in a data table you construct in your science notebook. Be sure to include comparisons of the differences in results obtained with different metals.
5. Using the plastic spoon, remove the pieces of metal from the cups and place them on a paper towel. Clean the spoon between each cup.
6. Make a record of your observations of the solutions left in each cup. Be sure to include comparisons of the differences in results.
7. Examine each of the metal pieces and describe what you observe. Be sure to include comparisons of the differences in results.
8. Using a dropper, transfer about 5 drops of each of the solutions to a clean cup in the tray. For example, transfer 5 drops from Cup 1 to Cup 5, from Cup 2 to Cup 6, etc. Clean the dropper between each cup.
9. Test for copper in the solutions by adding 2 drops of household ammonia solutions to Cups 5, 6, 7, and 8. Copper ions form a deep blue color or a blue-green precipitate in the presence of ammonia. Remember that Cup 8 contains the unreacted copper chloride solution (the control).
10. Record your observations. Your teacher will tell you how to dispose of each metal and clean out your tray. Complete the Analysis section on the next page.

ANALYSIS

Prepare a written report. Start your report on a clean sheet of paper. Include your name, the date, and a title for your report. Your report should have the following four components:

1. A statement of the problem you were trying to solve
2. A description of the materials and procedure you used to solve the problem
3. A clear presentation of the results you obtained
4. An analysis of the results. Include your conclusions, any problems you may have had with the investigation, and any additional questions related to the problem that you would like to investigate. Your analysis should include answers to the following two questions:
    - Overall, which metal seemed to work best at removing the copper from solution? Describe your evidence completely in your answer.
    - The table below summarizes typical information about the costs and legal levels of iron, aluminum, copper, and zinc ions that may be disposed of in wastewater. If you were to use one of these metals to remove the copper ions from the used solution, which one would you choose? Give reasons for your choice. Consider the results of your investigation as well as factors such as cost, allowable disposal levels in the wastewater, and possible uses of the metals for other purposes.

Information About Four Metals

Metal	Estimated Cost	Maximum Level in Waste Water	Availability
Aluminum	$0.90 per pound	not restricted	wide
Iron	$0.02 per pound	100 ppm	wide
Zinc	$0.59 per pound	5 ppm	wide
Copper	$1.33 per pound	5 ppm	wide

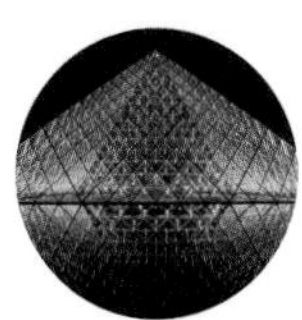

Put a Little Color Into Your Life

What do mercury, lead, chromium, copper, cadmium, and silver all have in common? They are all metals—heavy metals. They form colorful compounds, or colorants, that have been used since ancient times in paints, dyes, inks, and cosmetics. These metals have enriched and colored our lives. But if they are released into the environment carelessly, they can be toxic to living things.

CHALLENGE

Read about and discuss the historic uses of toxic heavy metals and their risks and benefits to society.

Malachite ore, an important source of copper

Copper: The Glory of Blue-Green!

Copper and its compounds have many beneficial uses. Copper plating helps protect metals from corrosion. Without wires made of copper metal, the long-distance transmission of electricity would be too expensive. Copper compounds provide the rich blue-green color of turquoise jewelry. The ancient Egyptians discovered that when malachite, a copper ore, was powdered and mixed with natural oils, it made an excellent cosmetic. The greenish-blue color was the perfect complement to women's eyes and was very fashionable. Cleopatra probably used it for eye shadow!

Copper in very low concentrations is essential for human health. Yet in slightly higher concentrations, copper compounds are toxic to living things. Before the U.S. Pure Food and Drug Act of 1906 was enacted, bluestone, or copper sulfate, was added to pickles to make them look fresh and have a dark green color. It was once commonly poured into public swimming pools to control algae. Today, copper compounds are no longer used in food, cosmetics, or pools.

Lead: From Pottery to the Color of the Queen's Face!

Lead is one of the densest of all metals. For thousands of years potters used lead compounds to create the colorful glazes on their pottery. Unfortunately, these compounds are easily leached by mild acid solutions and can be transferred directly from dishes and containers to the food they hold. Today, in the United States, pottery and ceramic items must pass rigorous tests to guarantee that they are essentially free of lead. Older ceramic items, especially imported ones, should be tested for lead before they are used for eating or drinking purposes.

Lead has no known function or health benefit for humans. Its compounds are highly toxic. They are easily absorbed through the skin, the lungs, and the digestive system. Because the body cannot tell the difference between lead and calcium, lead accumulates in bones and over time can build up to potentially toxic levels. Young children are especially susceptible to the toxic effects of lead when they eat chips of paint from older buildings that were painted before the early 1980s.

In the past, lead was not only used in glazes, it was also used in face powder, paints, inks, and even food coloring. In 16th century England, Queen Elizabeth I used a face powder made from white lead. The daily application of lead from the powder built up in her body, eventually leading to her death.

Until the early 1980s, lead made up nearly 50% of white house paint. Bright red, orange, and yellow paints were also colored from a mixture of lead and chromium compounds. Inks also contained lead. Red lead was used in food, too. Less than 100 years ago this red colorant was added to candy and tea as a food coloring.

Recently it was discovered that a new source of lead was entering the human diet. This time it came from the ink on plastic bread bags. It happened because some people who wanted to recycle the bags turned them inside out and reused them to store other food. Of course, any slightly acidic material, such as fruit, could cause the lead to leach from the ink. Researchers estimated that the lead leached was nearly double the normal intake of lead for most people. The researchers concluded that lead ink on such bags should be prohibited as an unnecessary health risk.

Queen Elizabeth I, of England

Coal and Petroleum: Colorful Synthetics

Both lead and copper compounds are regarded as natural colors. They are found in their natural state in the Earth as minerals, and they were used in much the same way as natural dyes extracted from plants. As we have seen, many of these natural mineral products were quite poisonous. However, there were no alternatives for these colorants until the discovery of petroleum and, with it, the discovery of synthetic colors and dyes.

In 1856, Sir William Henry Perkins of England synthesized the first artificial purple dye, mauve, from coal tar oil. Before that, purple dyes had come from mollusks and other natural sources. About three years later, the discovery of oil in Pennsylvania made large amounts of petroleum available, and other synthetic dyes and colorants were soon developed. These quickly proved to be superior to the plant and mineral dyes that had served civilization for more than 20 centuries. By 1900, a total of about 80 dyes had been developed.

Over time, with testing and strict governmental controls, many coal tar- and petroleum-based dyes have become a part of the inks, dyes, and food colorings we use today. They have replaced many of the early mineral pigments in paints and inks that were toxic to humans. Some recently developed inks are soy-based and are considered safer than the petroleum-based products. Today, all colorings must undergo rigorous tests before they are allowed to be used in the products and foods that we use daily.

First domestic discovery of oil, Western Pennsylvania, 1859

Reducing Hazardous Waste

For many years people have been using dyes to color clothing, paint, cosmetics, and other products. Have you ever wondered how the beautiful colors of some dyes are produced? In this activity you will learn about the materials used to produce dyes and some of the problems with disposing of these materials.

CHALLENGE

Compare the properties of two inks and consider the trade-offs involved in choosing one of them for printing a newspaper.

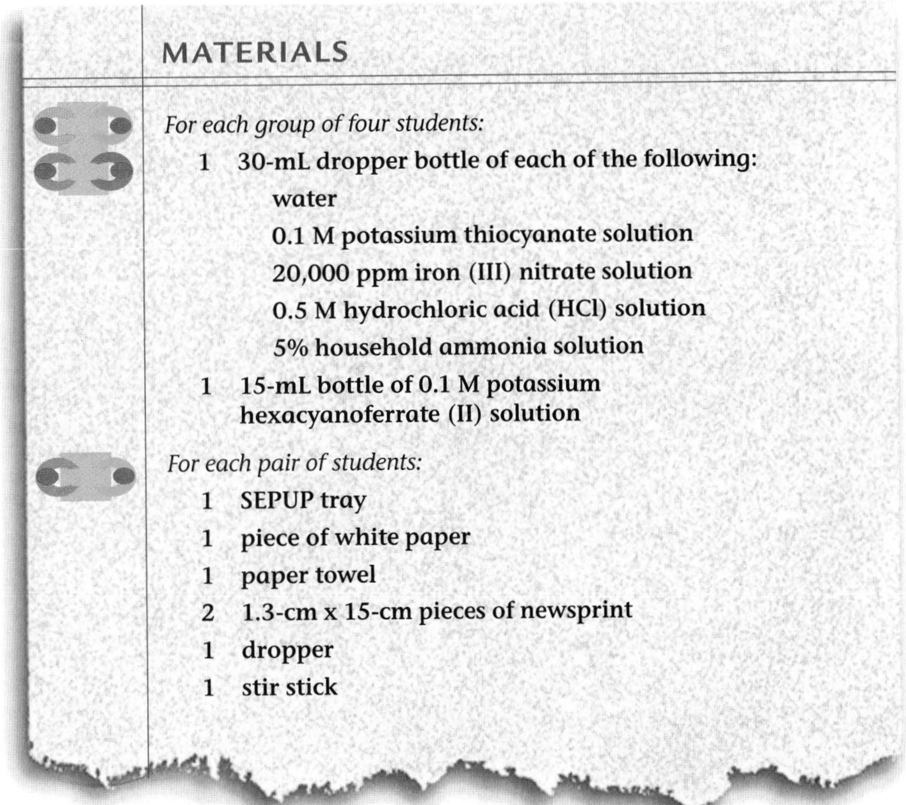

MATERIALS

For each group of four students:

1 30-mL dropper bottle of each of the following:
- water
- 0.1 M potassium thiocyanate solution
- 20,000 ppm iron (III) nitrate solution
- 0.5 M hydrochloric acid (HCl) solution
- 5% household ammonia solution

1 15-mL bottle of 0.1 M potassium hexacyanoferrate (II) solution

For each pair of students:

1 SEPUP tray
1 piece of white paper
1 paper towel
2 1.3-cm x 15-cm pieces of newsprint
1 dropper
1 stir stick

PROCEDURE

SAFETY NOTE: Wear protective eyewear. Caution! the inks in this experiments may permanently stain your clothes.

Part One: Testing for Iron

1. Take out your SEPUP tray. Fill Cups A–C three-quarters full of tap water.
2. Add one drop of 20,000 ppm iron (III) nitrate solution to Cups A–C. Stir.
3. Add 2 drops of household ammonia solution to Cup A. Stir. Clean the stir stick on the paper towel.
4. Add 2 drops of 0.1 M potassium hexacyanoferrate (II) solution to Cup B. Stir. Clean the stir stick.
5. Add 2 drops of potassium thiocyanate solution to Cup C. Stir.
6. In your science notebook, prepare a data table to record the solutions and colors in each cup.

ANALYSIS

1. What do the mixtures in all of the large cups have in common?
2. If the 20,000 ppm iron (III) nitrate solution represents a simulated toxic heavy metal, what concerns would you have about using these colored solutions for colorants in inks and paints?
3. What is the approximate iron concentration in Cup C? (**Hint:** See Activity 12.)

Part Two: Making Your Own Inks

Historically, the blue color you produced in Cup B has been used to make blue inks and dyes. It is called Prussian blue. In this part of the activity you will investigate two unknown blue ink samples. One ink represents Prussian blue and contains a simulated toxic heavy metal. The other ink represents modern replacements that have little or no heavy metal content. It will be your job to evaluate how useful each would be as a newspaper ink and recommend which ink to use.

1. Each pair of students in your group should test one of the ink samples —A or B. Decide which sample each pair will test.
2. Take your piece of newsprint to the stamp pad station you selected. Carefully apply the stamp to the ink and then press it to the paper. Repeat this three more times to produce a total of four equally spaced images on the piece of paper, as in the example below.

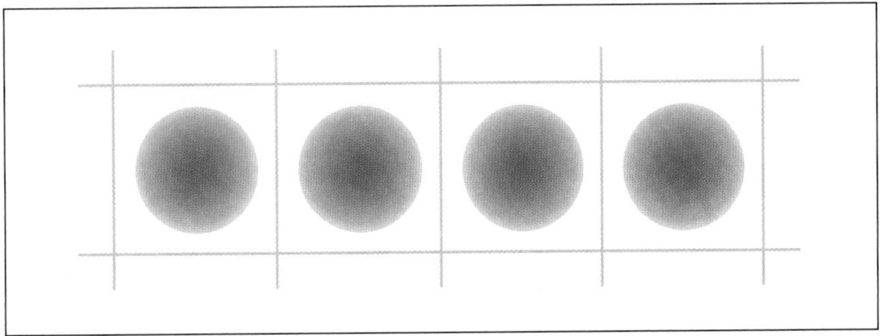

3. Cut the stamped pictures into four small, equal-sized pieces to fit the small cups of your SEPUP tray.
4. Each team will use one SEPUP tray to test its ink sample.
5. Put the pictures into small Cups 1–4. Wait 2 or 3 minutes for the ink to dry.
6. Add 6 drops of water to Cup 1.
7. Add 6 drops of bleach solution to Cup 2.
8. Add 6 drops of hydrochloric acid solution to Cup 3.
9. Use your dropper to add 6 drops of soap or detergent solution as assigned by your teacher to Cup 4. Wait 5 minutes.

10. Make a data table like the one below. Record your observations of the contents of each cup under the column for the ink sample that you are testing.

11. Next, in your data table, record the test results obtained by the other team in your group.

12. Add a drop of hydrochloric acid solution to Cup 1 and Cup 4. Remember that Cup 3 already contains acid. Add a drop of potassium thiocyanate solution to Cup 1, Cup 3, and Cup 4 to test for the presence of our simulated heavy metal (iron).

13. Record the results of your tests in a data table like the one on the next page.

14. Clean up, as directed by your teacher.

Comparison of Two Unknown Inks

Cup	Substance	Effect on Ink A	Effect on Ink B
1	water		
2	bleach		
3	hydrochloric acid		
4	soap/detergent		

Results of Testing an Ink for Simulated Heavy Metals

Cup	Color of solution with hydrochloric acid and potassium thiocyanate	Heavy metal present? Yes or no?
1		
2		
3		
4		

ANALYSIS

1. Which ink contained the simulated heavy metal?
2. What is the purpose of the water in Cup 1?
3. What evidence do you have to prove that you tested two completely different inks?
4. To be useful for printing, inks must not be easily washed out by water. Which ink is the most resistant to water?
5. Printing inks must not lose their color quickly when exposed to light. The bleach solution was used to represent the effects of sunlight bleaching. Which ink is the most resistant to bleaching?
6. Some inks, over time, react with the acids that are found in many papers. Which ink was the most affected by acid?

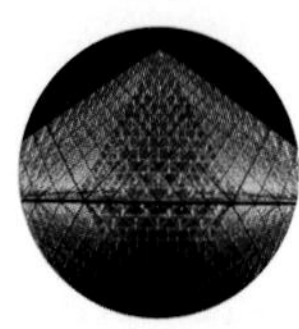

The Green Dot

Many local and national government agencies require their suppliers to use products that contain 25% recycled materials. Without these regulations, many recyclables would never be used because the raw materials that are used to make new ones are cheaper. To cut down on the amount of solid waste produced, governments in Europe require manufacturers to take back all their waste packaging. The following article will help you understand this program by showing you how a common product—an ink jet cartridge used in some computer printers—was redesigned to reduce the packaging and the added waste the manufacturer would be required to dispose of in a landfill or waste-to-energy incinerator.

After you read the article, think of ways to reduce the packaging on some products you use.

Part One: Redesigning Packaging

European countries were the first to employ large scale incineration of their household waste. People there wondered why so many things they used couldn't be reused, reduced, or recycled. They asked, "Why do so many things need to be thrown away?" One of the easiest areas to change (reduce) was packaging.

In the early 1990s, laws were passed to reduce packaging waste by giving products the "green dot" environmental symbol of approval. Green dots told consumers they could return the used packaging to the place they bought it or to a collection site, and the producer would be responsible for its reuse or disposal.

Manufacturers soon got the message. They began to redesign their products in order to reduce the amount of packaging used. For example, liquid detergents that were packaged in large plastic containers were replaced with a highly concentrated form of detergent in smaller paper containers. These concentrates could now be diluted in the larger plastic container, just as orange juice is made from a concentrate. The large container could be reused over and over again.

What follows is a case study of how a producer of ink jet printers in the United States, Hewlett-Packard Company, completely redesigned its packaging for its ink cartridges in order to reduce the waste and obtain the green dot in Europe.

Part Two: Packaging Ink Cartridges—A Case Study

Have you had a message from a computer lately? Computers can communicate with us through the monitor screen—and some can even talk. But to keep a written record of our work, we use a printer. A common and inexpensive kind of printer, called an ink jet printer, sprays dots of ink to form images. These printers are used in homes across the United States.

Ink jet printers use a removable, single-use cartridge that fits inside the printer. It moves back and forth across the paper surface, spraying out a specially engineered ink. The ink creates a changing pattern of dots that can become any style of type.

The ink jet cartridge head contains a gold-plated computer chip that determines which small holes the ink will flow through. Tiny, heat-producing resistors are connected to each hole and to a supply of ink. The computer sends a message code to the printer head causing the resistors to instantly heat up and produce a small ink bubble. The bubble explodes through the hole and sprays into the piece of paper, forming a dot. The pattern of dots produced by this and other exploding holes firing ink missiles at the paper forms the letters we eventually see.

The ink cartridge is a marvel of modern engineering techniques. But like any high-technology device it may encounter problems during shipping and storage before it ever reaches the consumer.

The inks in the cartridges are water based. They were designed to dry quickly, be nontoxic, and leave a lasting image on many different types of paper. Over time in storage, the water can evaporate, leaving a more concentrated ink behind—one that may clog the tiny holes in the cartridge's printer head. A second problem is transportation, specifically the bumps, shakes, and drops the cartridge must endure on the way to the consumer. These may also damage the delicate printer head. The original package was designed to prevent evaporation and to protect the cartridges.

The original cartridge package had seven separate components as shown below.

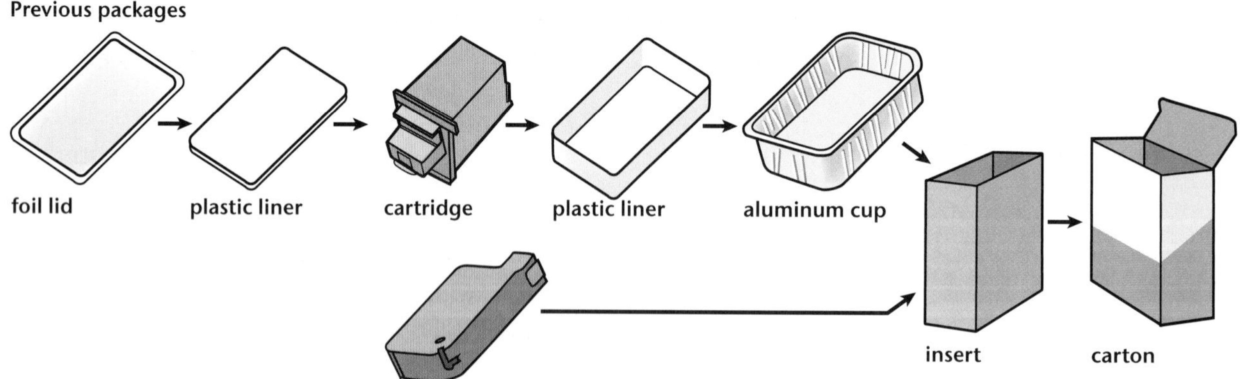

Aluminum was chosen because it prevents water from evaporating from the cartridge. The plastic liners and carton helped cushion the shocks of transport. But when this product was introduced into Europe, it had to reduce its packaging to qualify for the green dot.

This was a formidable challenge to the designers. After considering the trade-offs, both of cost and high quality for the consumer, they redesigned the whole cartridge. The aluminum barrier to prevent the ink from evaporating became an inside layer of the ink bag in the cartridge. Sturdy plastic parts were employed, and a carton with air cells (like the foam you studied in Activity 10) became a part of the packaging. What had been seven components, now became three. And Europe's gain became ours, too! The same cartridge is widely used here as well.

Manufacturers are now considering many ways to reduce the amount of packaging that eventually reaches landfills. Compact discs (CDs) once came in much large packages. New designs reducing the amount of plastic in soft drink bottles are being used today. As new materials and products appear on the store shelves, they will always need some form of packaging. Companies will be continually challenged to provide minimal packaging that reduces waste in order to be environmentally responsible.

QUESTIONS

1. What are at least five purposes that packaging serves?
2. Think of a product that you use frequently and how it is packaged. Name the product, give three reasons you think it is packaged this way, and make two recommendations for reducing the amount of the packaging materials.
3. Is it possible to sell products with no packaging? Explain.

Comparing Waste Management Plans

Some communities do not rely on one specific method for all types of waste. They use an integrated approach to waste management. This method brings together several approaches to create an overall plan to meet the needs of the community.

CHALLENGE

Your group will examine case studies of four counties across the United States that have developed waste management programs that outside groups have identified as outstanding.

PROCEDURE

Each group member should choose a different case study to read. When you finish reading, answer the questions below in your science notebook. Use the county name as the heading for your notebook entry.

QUESTIONS

1. What are the goals for managing waste in this county?
2. What methods are used to accomplish these goals?
3. What is unique about this plan? How does it take advantage of local conditions and contribute to protecting the quality of the environment?
4. What evidence is given about monitoring and safeguards to protect the environment from potentially toxic and hazardous substances?
5. What other information would you like to have about the county plan?

Waste-to-energy facilities like this one convert chemical energy in waste to useful forms such as electricity

Case Study 1: Mecklenburg County, North Carolina, 2002

Spurred by the economic development of its principal city, Charlotte, Mecklenburg County is a growing area. The current population is expected to increase by 30% by the year 2006, and employment opportunities will increase by 50%.

In 1985, most of the county's waste was sent to landfills. County officials then began to develop a plan for integrated solid waste management. Current elements of the plan include a municipal solid waste (MSW) landfill, various specialty sites, medical waste incinerators, composting facilities, and curbside collection facilities. Today, roughly 36% of the County's measurable waste stream is disposed of or reclaimed for use within its borders; the rest is exported. Despite a 3% growth in population in 2000–01, the amount of MSW dropped by 4%.

With the use of color-coded containers, curbside collection of aluminum, glass, newspaper, and plastic began in all towns and cities within the county. Now, over 74% of homes participate in curbside recycling. Nearly one-fourth of all yard waste—leaves, grass, and other clippings—is recycled and composted. An extensive recycling program for commercial waste to recover office paper, corrugated cardboard, iron-containing metals, and scrap wood will begin soon.

Municipal Solid Waste, 2000–01	
Municipal solid waste	62%
Household hazardous waste	1%
Tires	1%
Yard waste	4%
Construction and demolition	31%
Waste water sludge	2%
Other, includes regulated medical waste, etc.	<1%

In 1989, the county's first waste-to-energy incinerator began operation. Approximately 10% of the county's solid waste is treated in this incinerator. Steam from the facility is sold during the winter months to heat university campus buildings. Excess electricity is sold to utilities. Total annual energy earnings are estimated at $0.8 million, while the county must spend $2.4 million yearly to operate the plant.

Adapted from *Case Studies in Integrated Waste Management.* Used with permission by the Council on Plastics and Packaging and the Environment.

Charlotte

Case Study 2: Pinellas County, Florida, 2002

Surrounded on three sides by the Gulf of Mexico and Tampa Bay, Pinellas County offers a picturesque environment. But county solid waste authorities must also contend with nearly 25,000 new residents each year and a waste generation rate that is nearly 75% higher than the national average. Pinellas County is the most densely populated county in the state, with over 3,111 people per square mile. To address its solid waste disposal needs, the county developed a four-part program that has helped reduce its reliance on landfills by 89% since 1983. The county's plan includes recycling, composting, a modern waste-to-energy plant, and an artificial reef program.

Recycling is an important part of Pinellas County's overall waste management strategy. A coordinated program was begun in 1989 in response to legislation passed in 1988. Pinellas County reached a 30% recycling goal and has sustained and surpassed that rate since 1992. Components of the plan include curbside collection and drop off centers for office paper, cardboard, aluminum foil, and used motor oil.

Pinellas County's waste-to-energy plant began operations in 1983. The facility was built to handle 2,100 tons of municipal solid waste per day. A new boiler, added in 1986, increased the processing capacity of the plant to 3,150 tons per day. It operates year round, 24 hours a day, and is one of the largest facilities of its kind in the country. Its environmental impact is carefully controlled and monitored by engineers, and air emissions from the plant are regularly tested. Most of the county's solid waste is burned to produce electricity and the resulting power is sold to the local utility. All metals are extracted from the ash and sold as scrap for reuse. The residue is a non-hazardous aggregate material composed of sand, ash and pulverized glass. The material is used at the Pinellas County Solid Waste facility as soil substitute, landfill cover, road base, and for other practical, cost-efficient uses.

Before Pinellas County's waste-to-energy system was implemented, 100% of the waste generated went to landfills. This figure has dropped dramatically.

Only 7% of the country's solid waste now goes into the one remaining landfill. This reduction has greatly lengthened its life. The entire solid waste system was designed to be self-supporting and receives no funds from the county tax base. Its operating income is derived from user fees and the sale of electricity and recovered metals.

Pinellas County maintains one of the largest artificial reef programs in the country. It reduces the impact of large, inert material in the landfill and increases the fish population in the Gulf of Mexico. The artificial reef program uses large volumes of concrete debris to build underwater marine habitats. Concrete culverts, cut-off pilings, storm drain junction boxes and seawall slabs are some of the materials that have been used to build ten artificial reefs that attract a variety of sea life.

Adapted from *Case Studies in Integrated Waste Management.* Used with permission by the Council on Plastics and Packaging and the Environment.

Tampa

Case Study 3: Hennepin County, Minnesota, 2002

Hennepin County, which includes the city of Minneapolis and outlying communities, has developed an integrated approach to its waste management needs. The county is home to over 1.16 million people (2000 census data). Through county grants, widespread industry support, and hard work, the 46 cities that make up Hennepin County were able to reduce the amount of waste sent to landfills to 20% of total waste, down from 67% in 1989.

A permanent site for the collection of household hazardous waste was established. The county also began a program to collect batteries. There are more than 500 drop-off locations where batteries are collected and sent to a processor that reclaims the mercury and silver they contain.

Minneapolis

From 1995–1999, the Hennepin County population grew by 8.6% while the garbage rate grew by 20%. All cities within the county have curbside recycling programs. Major recyclables include glass containers, aluminum, tin (bimetal) cans, newspaper, cardboard, and plastic bottles. There is one county drop-off center plus 30 local centers. The county provides financial assistance for programs that separate recyclable items into categories at their source of collection. With vigorous effort, Hennepin County's recycling rate grew from 22% of solid waste in 1984 to a high of 50% in 1995. Unfortunately, in the past few years, waste generation has outpaced recycling growth. As of 1999, the most recent data available, the recycling rate was 47%.

Hennepin is the sponsor of a waste-to-energy plant located near downtown Minneapolis. The plant began operation in late 1989 and burns an average of 1,000 tons of waste every day, generating enough electricity for 40,000 homes. Part of the fees charged to the private and public waste haulers that use the plant support other waste management programs. A second waste-to-energy plant is the Elk River refuse-derived fuel system in Anoka County. This operation began in 1989 and handles up to 1,500 tons of waste per day from five counties, including Hennepin. The waste is turned into fuel and then burned by a local power company to generate electricity. Hennepin County has a contract with the Elk River plant to burn 200–800 tons of waste per day, depending upon the seasonal flow of refuse.

Presently, 53% of Hennepin County's waste is processed by the waste-to-energy plants. These plants use dry scrubbers and baghouses as pollution control devices. They operate under the limits set by the State Pollution Control Agency.

Adapted from *Case Studies in Integrated Waste Management.* Used with permission by the Council on Plastics and Packaging and the Environment.

Case Study 4: Marion County, Oregon, 2002

Like many communities, Marion County used landfills for all of its waste disposal for many years. With the closure of the Brown's Island landfill in 1974, Marion County's 284,000 residents were left without an immediate disposal method. Through a comprehensive waste management program, the county was able to reduce the amount of waste sent to landfills from 100% in 1974 to just 7% of total waste today.

In 1971, Oregon became the first state to pass bottle deposit legislation. The law places a deposit on all soft drink and beer containers and requires customers to return the containers to receive a refund. Oregon also has mandatory recycling for all cities with a population of more than 4,000.

The county went a step further in 1986, when it began a curbside recycling program. It requires trash haulers to pick up recyclables (glass, newspaper, aluminum, tin cans, corrugated cardboard, and used motor oil) on a weekly basis in urban areas and on an on-call basis in rural areas. There are also 20 drop-off locations in the county that accept these same materials along with grayboard, magazines, and PETE and HDPE plastic (recycling code 1 or 2).

Silver Falls, Marion County

The program is funded by the disposal and collection fees paid by the haulers. Haulers also pay for the drop-off program.

A decline in the scrap value of many recyclables has led to a $90,000 deficit for the recycling program. Currently 25% of Marion County's municipal solid waste is recycled. Yard waste accounts for more than 20% of the county's municipal solid waste. There are four yard-waste composting facilities operating in the county. The county charges $55 per ton, and the two private facilities charge $40 per ton. This pays for the operation of the program. Approximately 10% of Marion County's yard waste is composted. The other new materials now collected are computer components, which began in July, 2002. This has been extremely well-received, and in less than 6 months, about 93 tons were recovered. The computers are taken by a company that funnels the reusable computer equipment into the Oregon schools, salvages other usable components, and then sends the rest out for materials recovery.

Marion County is the sponsor of a mass burn (all municipal wastes collected are burned together) waste-to-energy incinerator. The plant was completed in 1986 at a cost of $47.5 million. The plant can process 170,000 tons per year. The plant's 13.1-megawatt generator produces 11 megawatts of electricity that are sold to a local power company for approximately $4 million a year. The pollution control devices for the plant are dry gas scrubbers and fabric filter baghouses. Ash from the plant is sent to a special landfill. Today, 66% of Marion County's municipal solid waste is burned to recover energy.

Adapted from *Case Studies in Integrated Waste Management.* Used with permission by the Council on Plastics and Packaging and Environment.

Preparing an Integrated Waste Management Plan

Imagine that the solid waste landfill site owned and operated by your community will be full in four to five years at the current rate of waste disposal. This is not enough time to locate a new site and build a landfill that meets state and federal environmental regulations. In order to extend the life of the landfill and allow time for a new one to be built, local government has mandated a 25% reduction in the amount of waste sent to the landfill.

CHALLENGE

Prepare an integrated waste management plan that will reduce the amount of waste sent to the landfill and extend the life of the landfill.

ADDITIONAL INFORMATION

To accomplish the challenge you should consider using a combination of waste management methods. Your plan should ensure that all wastes are reduced in volume and toxic content, and that useful energy and materials are extracted prior to final disposal. Remember, even when the new landfill is built, it won't last forever, and you want your community's long-term waste management plan to be safe and effective.

Past recycling efforts in the city have been sporadic and small in scale. There is no curbside pickup, and there are no neighborhood collection sites. The city does not operate a waste incinerator. The percent by volume and weight of the city's solid waste is provided in the pie charts on the next page. Keep in mind that almost one-third of the paper in the landfill is newsprint.

Composition of Municipal Solid Waste by Weight, 2000

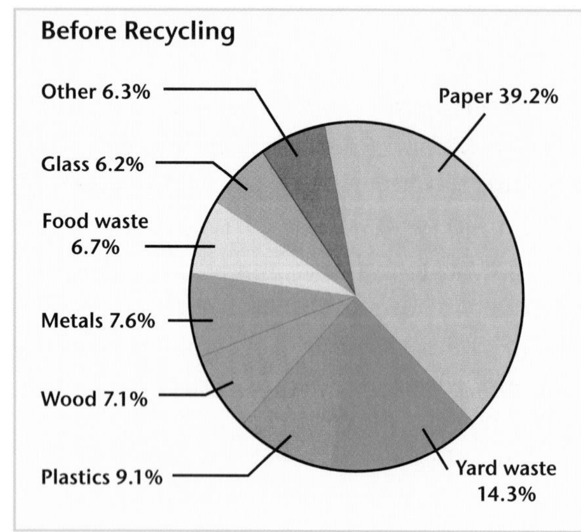

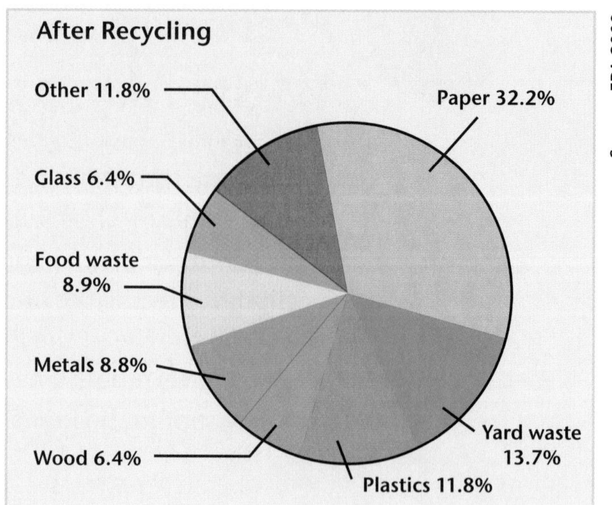

Composition of Municipal Solid Waste by Weight, 2000

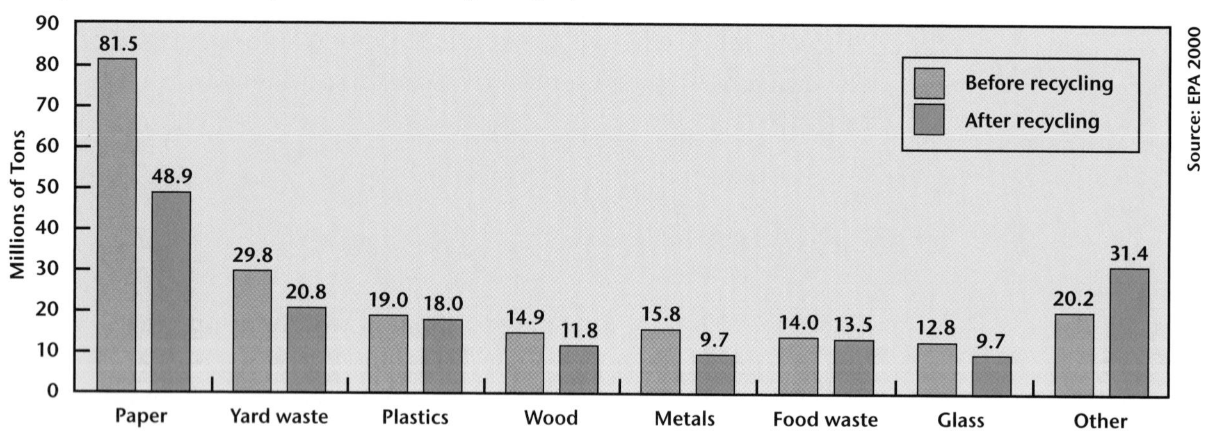

Total and Net Waste Generation, 1960 to 2000

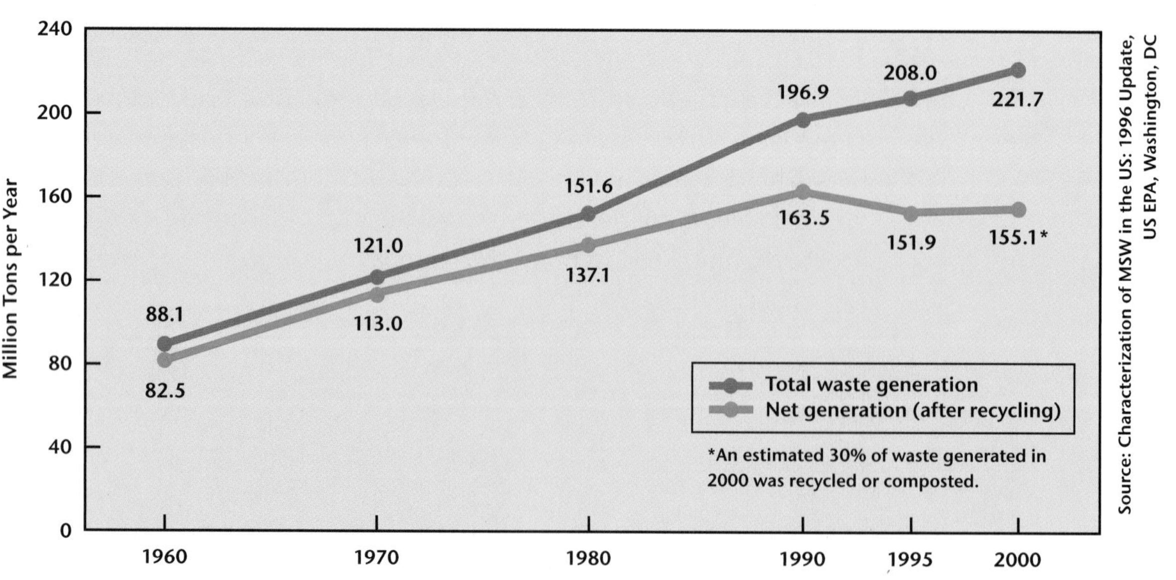

PROCEDURE

1. Working in groups of four, develop an integrated waste management plan for the community. You may find it helpful to consider each category of waste described on the pie graphs on the previous page. Work with your group to identify at least five items that you think should be included in the waste management plan. Prepare a table to summarize your ideas.
2. Decide how your group will present its solution to the class. In your presentation, explain why you are convinced that your plan will accomplish the 25% reduction in the amount of waste going to the landfill for disposal.
3. In your science notebook explain the highlights of your group's plan and how the plan will reduce the waste going to the landfill by 25%.

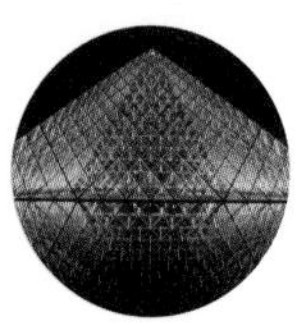

Materials and Applications

MAKING NEW MATERIALS

The materials we use to make things have changed over time. From simple materials, like stone, wood and glass, we now have an almost bewildering variety of new materials to choose from. Keeping track of when new materials—and new inventions—first appeared is important, not only from an historical perspective, but to understand how different technologies "build on" off one another.

For example, thousands of years ago, humans used stone tools. Over time, metals were extracted from the earth and used to make tools—first bronze (a mixture of copper and tin) and later, iron. The particular metal was even used to define the period—the "Bronze Age" and even the "Iron Age." Each advance was built on previous knowledge and experience.

MATERIALS THROUGH TIME

The last few hundred years have seen enormous changes in the human condition. World population has increased from less than 500 million people in the year 1400 to over six billion today. And the increase in the types and use of materials is just as spectacular. Technologies we take for granted today would have seemed like science fiction to our great-great grandparents.

CHALLENGE

Use the Materials and Application timeline on the next to look for patterns in human population growth and technological progress.

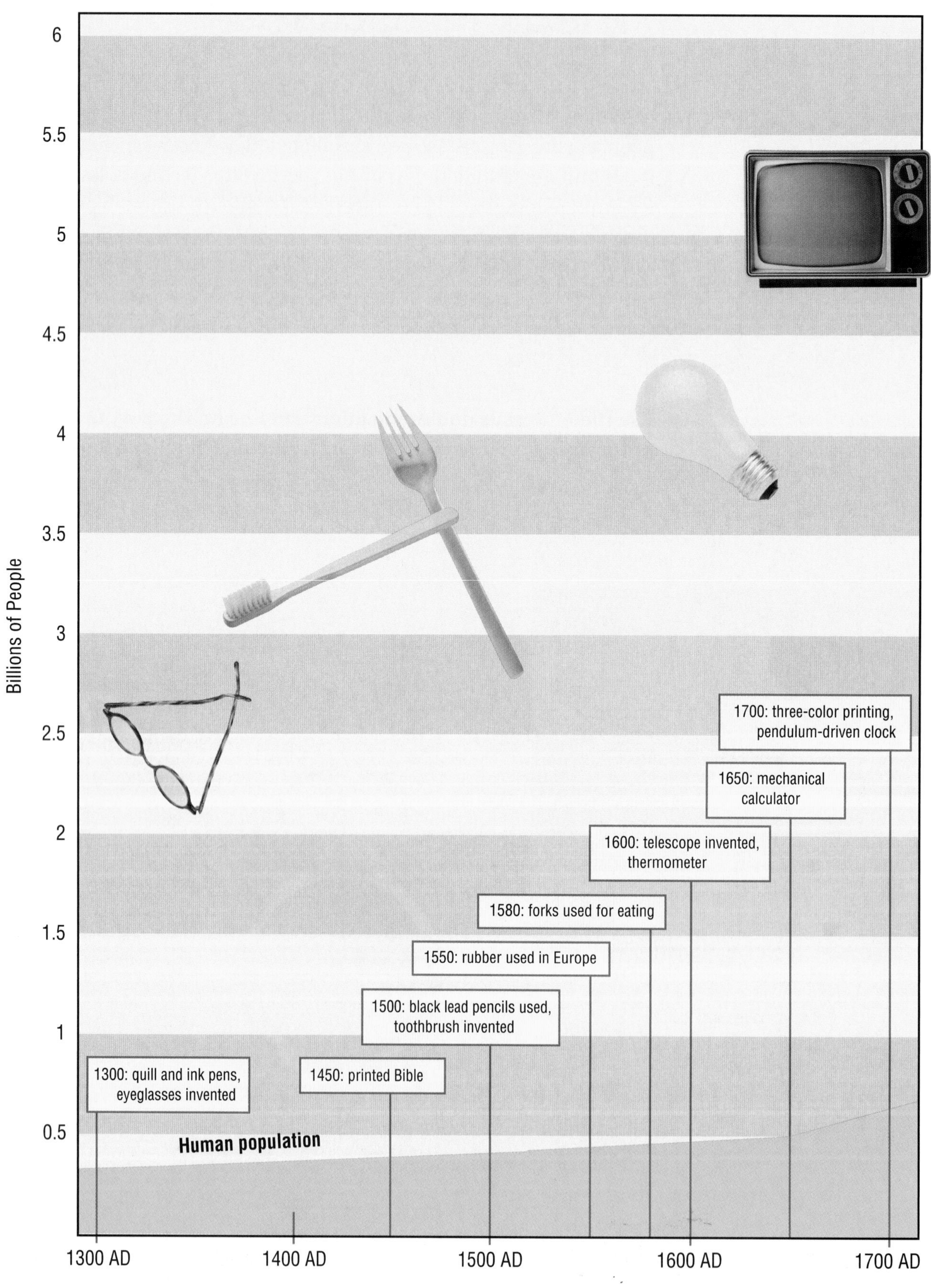
6
5.5
5
4.5
4
3.5
3
2.5
2
1.5
1
0.5
Billions of People
1300: quill and ink pens, eyeglasses invented
1450: printed Bible
1500: black lead pencils used, toothbrush invented
1550: rubber used in Europe
1580: forks used for eating
1600: telescope invented, thermometer
1650: mechanical calculator
1700: three-color printing, pendulum-driven clock
Human population
1300 AD
1400 AD
1500 AD
1600 AD
1700 AD

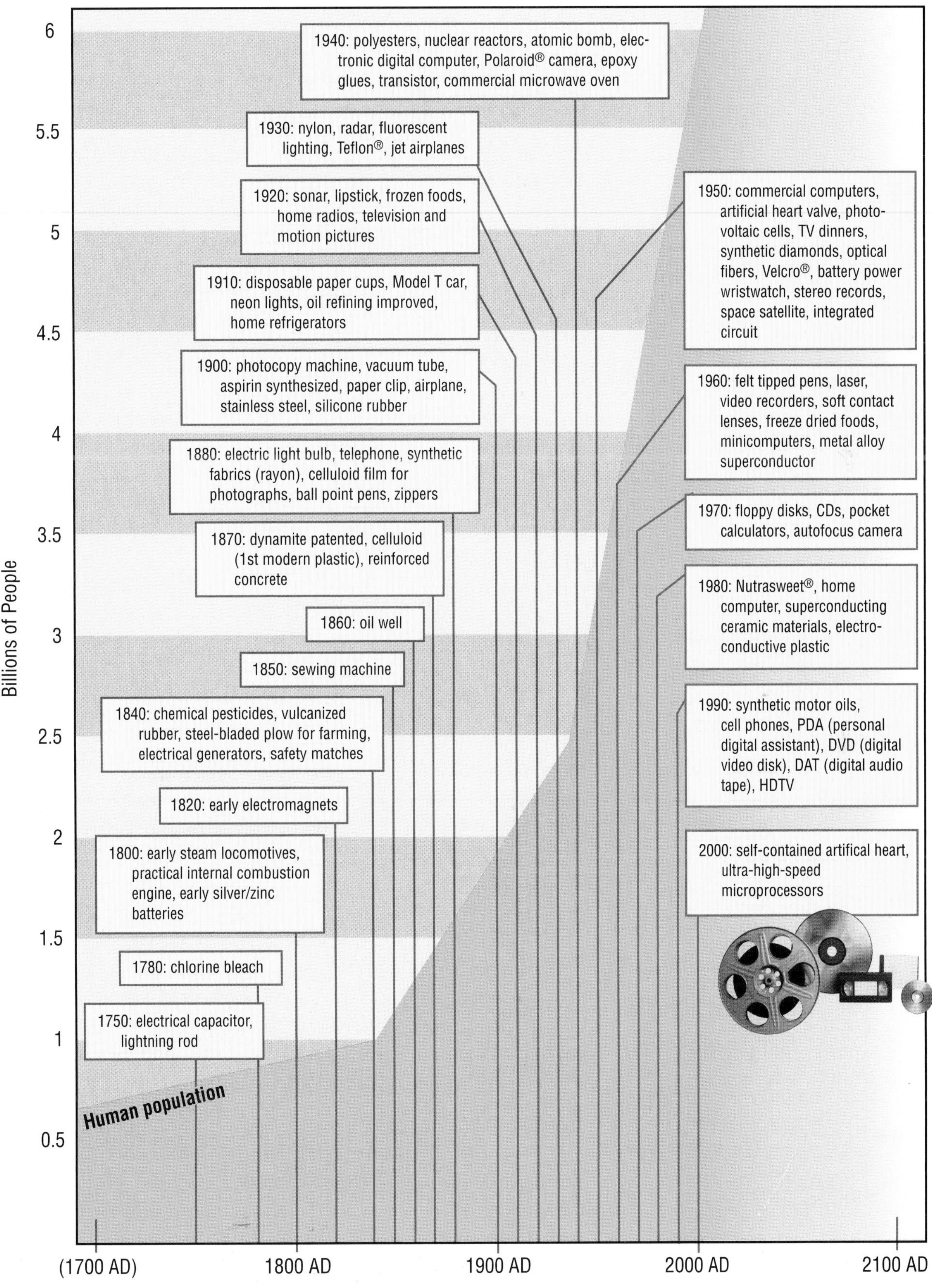

Billions of People
6
5.5
5
4.5
4
3.5
3
2.5
2
1.5
1
0.5
(1700 AD)
1800 AD
1900 AD
2000 AD
2100 AD
Human population
1750: electrical capacitor, lightning rod
1780: chlorine bleach
1800: early steam locomotives, practical internal combustion engine, early silver/zinc batteries
1820: early electromagnets
1840: chemical pesticides, vulcanized rubber, steel-bladed plow for farming, electrical generators, safety matches
1850: sewing machine
1860: oil well
1870: dynamite patented, celluloid (1st modern plastic), reinforced concrete
1880: electric light bulb, telephone, synthetic fabrics (rayon), celluloid film for photographs, ball point pens, zippers
1900: photocopy machine, vacuum tube, aspirin synthesized, paper clip, airplane, stainless steel, silicone rubber
1910: disposable paper cups, Model T car, neon lights, oil refining improved, home refrigerators
1920: sonar, lipstick, frozen foods, home radios, television and motion pictures
1930: nylon, radar, fluorescent lighting, Teflon®, jet airplanes
1940: polyesters, nuclear reactors, atomic bomb, electronic digital computer, Polaroid® camera, epoxy glues, transistor, commercial microwave oven
1950: commercial computers, artificial heart valve, photovoltaic cells, TV dinners, synthetic diamonds, optical fibers, Velcro®, battery power wristwatch, stereo records, space satellite, integrated circuit
1960: felt tipped pens, laser, video recorders, soft contact lenses, freeze dried foods, minicomputers, metal alloy superconductor
1970: floppy disks, CDs, pocket calculators, autofocus camera
1980: Nutrasweet®, home computer, superconducting ceramic materials, electroconductive plastic
1990: synthetic motor oils, cell phones, PDA (personal digital assistant), DVD (digital video disk), DAT (digital audio tape), HDTV
2000: self-contained artifical heart, ultra-high-speed microprocessors

QUESTIONS

1. List three examples of products that have changed as a result of the development of new materials.
2. If the rate of population growth stays the same, what will the world population be by 2010 AD? By 2050 AD?
3. In the past few hundred years, the world population has doubled again and again. This is called exponential growth. What has made it possible to sustain so many people?
4. What materials or material applications have benefitted people the most?
5. Why have materials and material waste issues become so important in society?

Energy

Energy

What is energy? How do we use it? Where does it come from? Are there enough energy resources for us to continue using energy at the current rate during the 21st century? How can we lower the amount of pollution we cause when we use energy? These are some of the questions you will investigate in this book.

Our personal energy levels often affect our choice of words. If we are too tired to perform a household chore or task, we may say we "don't have the energy" and need to eat or rest "to recharge our batteries." But energy comes in a variety of forms. The energy we get from food is different than the energy from fossil fuels, flowing water, and other sources we use to power our vehicles and other machines, plus our streets, towns, and cities. All these energy needs involve transformations (changing one type of energy for another) and have different environmental impacts. And although it might not seem so at first, our societal energy demands are really the sum of all of our individual needs. When you look at it that way, your own energy decisions are some of the most important decisions you make in life.

The activities in this book will help you learn more about energy sources. You will also learn how we use one type of energy to produce other types. (Think of burning fuel oil to make heat to make steam to power a turbine that generates electricity.) You will investigate energy chains and energy efficiency. Most important, you will relate what you learn about energy to your own life and think about yourself as an energy consumer.

When you have completed the activities in this book you will be able to use what you learn about energy and energy sources to come up with your own plans for a more energy-efficient future.

Build a House of Cards

It takes energy to construct a building. Where does the energy go? You will investigate that question in this activity.

Working as a group, you will have 10 minutes to build a house, up to five stories high, out of cards. Each card represents the amount of energy you use to lift it into place when you construct that part of the wall or ceiling. The house must be able to support a foam drinking cup on top. The house that has the greatest number of stories, uses the fewest number of cards, and stores the least amount of energy is considered the most successful.

CHALLENGE

Use the smallest amount of stored energy you can to build a five-story "House of Cards" that can support a foam cup.

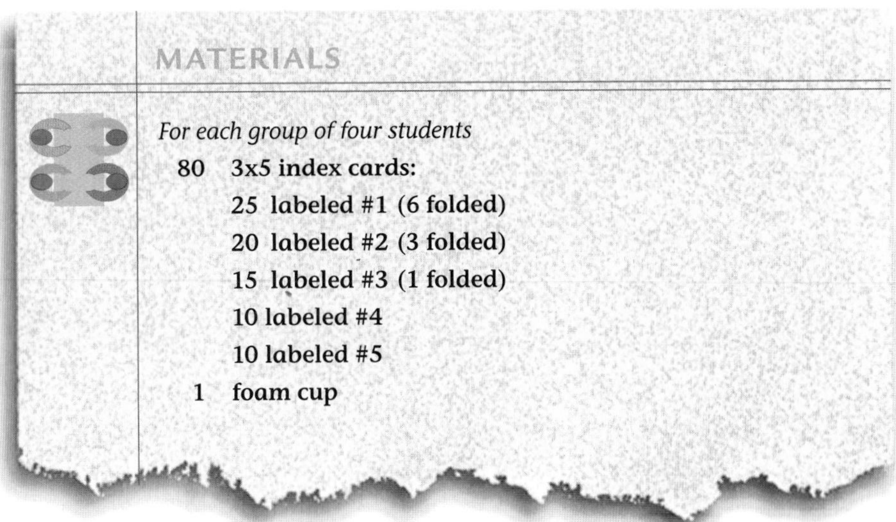

MATERIALS

For each group of four students

80 **3x5 index cards:**

- 25 labeled #1 (6 folded)
- 20 labeled #2 (3 folded)
- 15 labeled #3 (1 folded)
- 10 labeled #4
- 10 labeled #5

1 **foam cup**

PROCEDURE

1. Each group may build only one card house.
2. The structure must be freestanding and support the cup on top.
3. The cards may not be folded, except for the 10 prefolded ones.
4. The cards may not be ripped or cut.
5. Each "story" is made up of its walls and the ceiling above it.
6. Numbered cards correspond to each story and may be used only for that story.
7. You are allowed only 10 minutes for official construction of your final model. When the time is up, the structure must stand, with the cup on top, without collapsing for at least 1 minute.
8. When directed by the teacher, your group will release the energy now stored in the structure and calculate the number of "energy units" the house contained. Use the directions on the next page for calculating energy units.

ANALYSIS

Adding up the energy that was stored in your finished structure is easy. Because each card is the same size and weight (whether used as a wall or a ceiling panel), the energy it takes to lift one is the same for each. It's also easy to understand that lifting one twice as high takes twice the energy, and three times as high takes three times the energy, etc.

Energy cost per card:

Story 1 = 1 energy unit

Story 2 = 2 energy units

Story 3 = 3 energy units

Story 4 = 4 energy units

Story 5 = 5 energy units

Use a chart like the following one, or one your group designs, to help you calculate the stored energy:

Calculating Stored Energy Units

Story	Number of Cards	Energy Unit per Card	Total Energy Units per Floor
1		times 1	=
2		times 2	=
3		times 3	=
4		times 4	=
5		times 5	=

Drive A Nail

In this activity, you will design investigations to help you understand the differences between potential (energy of position) and kinetic (energy of motion) forms of energy. You will investigate the variables that determine how far a nail is driven into a foam block by a falling weight.

CHALLENGE

Design and carry out investigations to determine the effect of height and mass (weight) on the energy transferred to a falling object.

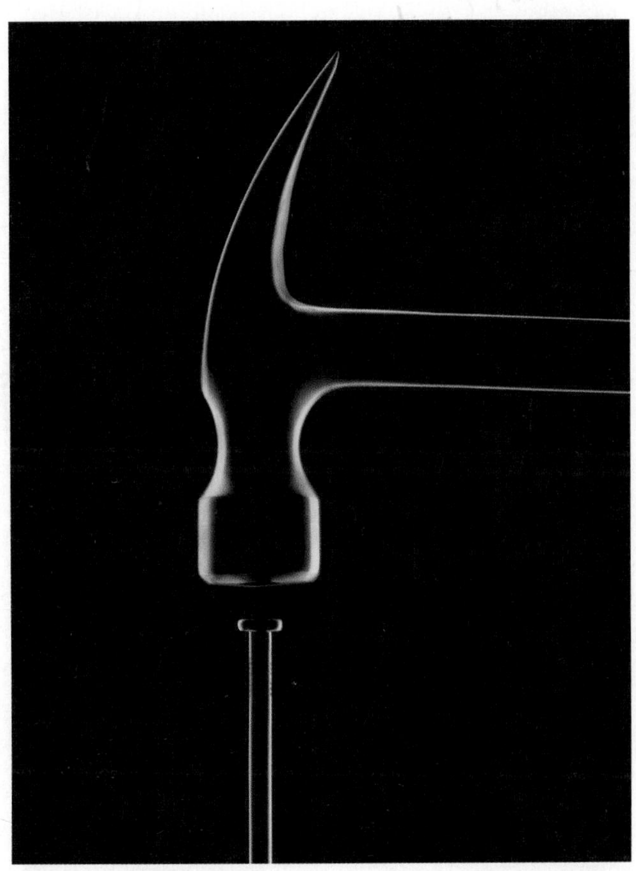

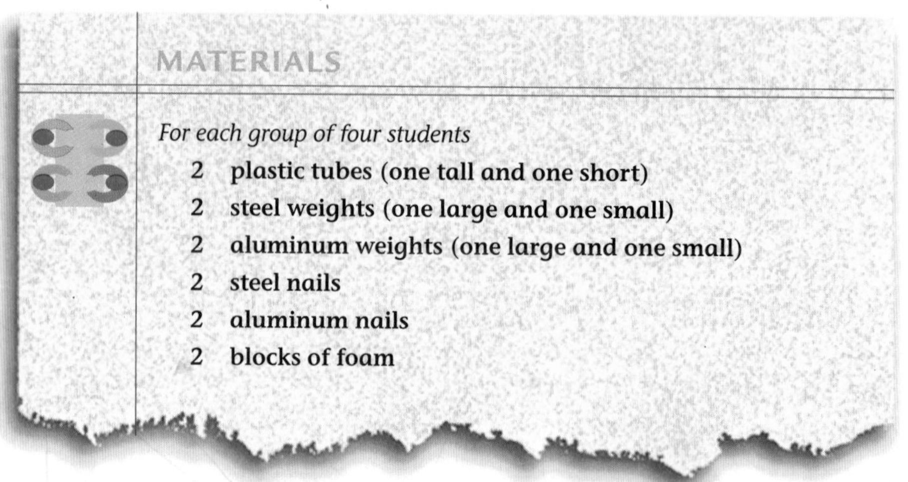

MATERIALS

For each group of four students

2 plastic tubes (one tall and one short)
2 steel weights (one large and one small)
2 aluminum weights (one large and one small)
2 steel nails
2 aluminum nails
2 blocks of foam

SAFETY NOTE: Use the materials your teacher provides. Follow the safety precautions demonstrated in the basic test procedure that your teacher modeled for you.

PROCEDURE

You have two plastic tubes—one twice as tall as the other—that will help you control the drop height. You also have two steel weights—one twice as heavy as the other—and two aluminum weights—again, one twice as heavy as the other. These will make it easy to control and keep track of the weight of the object you drop. This, in turn, makes it easy to investigate the effect of weight on the amount of energy transferred to the nail.

The actual testing will be done in the next science class. Take the remainder of this period to develop a group plan for that test. Write out the procedures your group will use. This will help you control all the variables except the one you want to change in each test.

Keep in mind: You want to carry out a controlled investigation that will determine the effect of drop height and weight on driving a nail completely into the foam. (Two members of your group might test the aluminum weights and two the steel weights.) Be sure everyone in your group is clear on exactly how to drop the weights from the top of the tubes. Any variation in method is likely to cause mixed results. Making a group data table now will help your group think about what needs to be tested and who is going to do what.

Before beginning your investigations, answer the following questions in your science notebook and explain your predictions:

1. Which combination of tube height (tall or short) and metal weight (steel or aluminum/large or small) will drive the nail deepest into the foam block? Explain how you arrived at your answer.
2. Which combination of tube height and metal weight will require the greatest number of drops to pound the nail completely into the foam block? Explain how you arrived at your answer.

ANALYSIS

Once you have performed your investigations, choose one of the combinations you investigated. In your science notebook describe the energy changes from potential to kinetic for the weight dropped. When was the potential energy at a maximum? When was the kinetic energy at a maximum? Explain.

GOING FURTHER

Ask your teacher for a different type of nail. Determine if there are any differences in the energy it takes to drive it into the foam block.

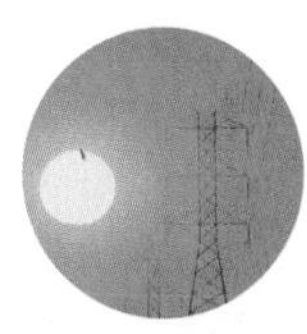

The Ice-Melting Race

This activity will help you understand some basic principles about the energy transfers that occur when we produce and use energy. Think about how the electrical energy that powers things such as lights, refrigerators, ovens, and computers gets to your home or school. It has to be generated in a power plant and transferred to your home. There, various electrical devices transform the electricity into useful forms of energy, such as heat or light.

Before refrigerators, ice was hand-delivered to homes.

In this activity you will investigate the direct transfer of heat energy from your body to an ice cube.

Melt as much ice as possible from an ice cube in just five minutes.

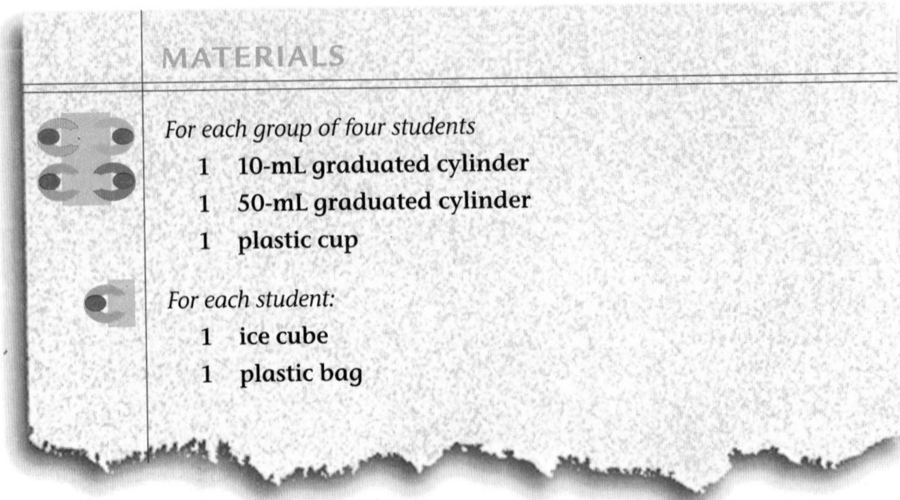

MATERIALS

For each group of four students

1 10-mL graduated cylinder
1 50-mL graduated cylinder
1 plastic cup

For each student:

1 ice cube
1 plastic bag

SAFETY NOTE: Do not put the bag in your mouth or under your clothes.

PROCEDURE

1. Get a plastic bag containing an ice cube. After you receive the bag, do not leave your seat. Do not remove the ice cube from the bag. Failure to follow these instructions will disqualify you from the race. Handle the plastic bag carefully. Breaking the bag also disqualifies you.

2. Try to melt as much ice as you can in 5 minutes. Your teacher will tell you when to start.

   Remember: Stay in your seat during the race.

3. When your teacher tells you that the time is up, carefully remove the unmelted ice from the bag and discard it in the cup provided. Carefully pour the water in your plastic bag into the large or small graduated cylinder and determine how much ice melted. Then discard the melted water so that another student can use the graduated cylinder. Answer the following questions in your science notebook.

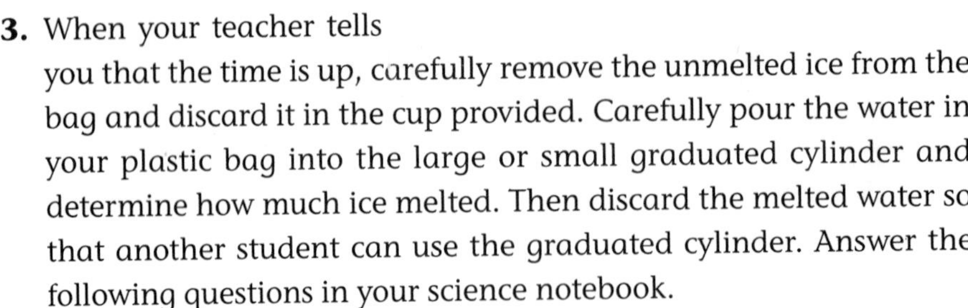

ANALYSIS

1. How many mL of ice were you able to melt in 5 minutes?
2. What energy sources did you use to melt your ice cube?
3. In what ways did you transfer energy in your approach to ice melting?
4. What energy transformations, such as mechanical energy [rubbing] to heat, took place during your participation in the ice melting race?
5. What did you do to increase the rate at which your ice melted? Did other students use different techniques? Compare the rate at which your ice cube melted with the melting rates of other students' cubes, and explain why they may have been different. Be sure you use your knowledge of energy and experimental design in your answer.
6. Describe all of the variables that affected how fast the ice melts. Choose one of these variables only and describe an investigation that would give a fair test of how changing this variable would affect the ice melting.

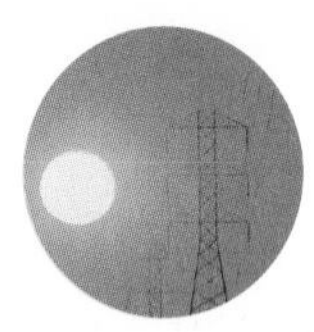

The Ice-Preserving Contest

In the ice-melting race, you tried to maximize the transfer of heat to melt the ice. In this activity you will do the opposite: try to prevent the flow of heat to the ice so that the ice will not melt.

Work with a partner to preserve an ice cube for as long as possible.

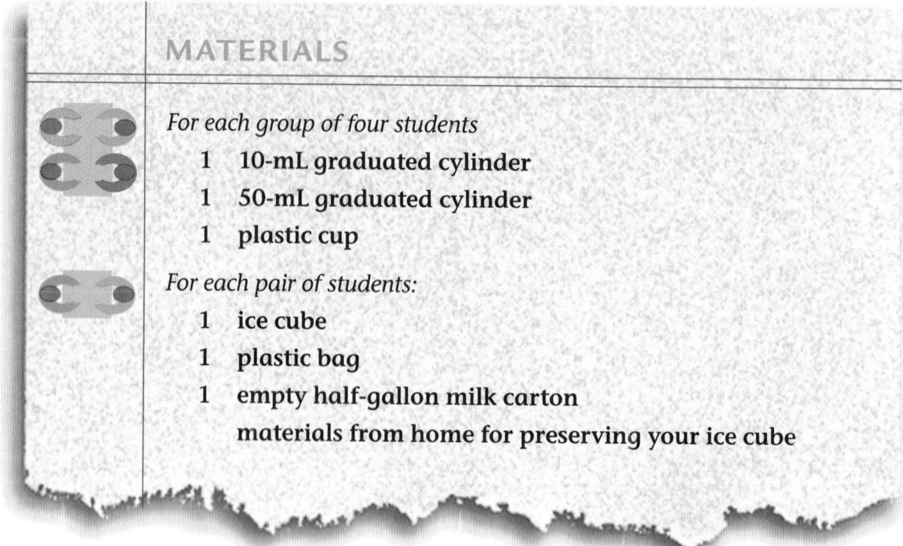

MATERIALS

For each group of four students

1 **10-mL graduated cylinder**
1 **50-mL graduated cylinder**
1 **plastic cup**

For each pair of students:

1 **ice cube**
1 **plastic bag**
1 **empty half-gallon milk carton**
materials from home for preserving your ice cube

PROCEDURE

1. Set up your ice-preserving investigation in the container provided. Allow it to remain undisturbed until your teacher tells you that time is up. You will be disqualified if your plastic bag with the ice cube leaks.

2. Carefully pour the water melted from the ice into the large or small graduated cylinder. Record the volume of water you obtained.

ANALYSIS

1. Describe what you did to preserve your ice cube. Explain why you thought it would work.

2. How much water melted from your ice cube by the end of the investigation?

3. What would you do differently in another ice-preserving contest?

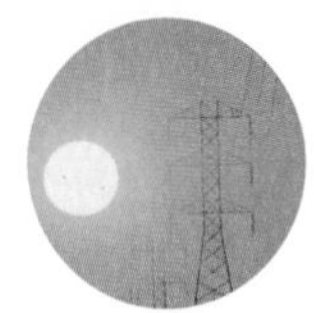

Refrigerators, Iceboxes, and Heat Transfer

Keeping our homes comfortable and our food from spoiling involves the transfer of heat. In some cases we want to encourage heat transfer, while in other cases, we want to prevent it.

CHALLENGE

Think of old-fashioned iceboxes and modern refrigerators in terms of desirable and undesirable transfers of heat energy as you answer the questions that follow the reading.

KEEPING FOOD COLD THROUGH THE AGES

It's the hottest day of the summer. Your friend comes to visit, and the two of you decide to go outside and relax. On your way outside, you stop in the kitchen, open the refrigerator, and find some ice-cold sodas. You take the sodas outside and alternate between sipping your cool drink and holding it against your hot forehead. Some of the heat from your body transfers to the cool drink. It cools your body and feels great.

Living in the 21st century has some distinct advantages, including the availability of modern refrigerators. The refrigerator is America's most common appliance. It is found in more than 99.5% of all homes in the United States. Refrigeration helps us to enjoy cool drinks on hot days. More important, it helps prevent food from spoiling.

Spoiled food has always been a health risk. In the past, many more people got sick or even died from food poisoning than today. However, even today, public health officials believe that millions of people suffer from food poisoning in the United States each year. While most recover in a day or two, some people, especially children, the elderly, or those weakened by other illnesses, die as a result of food poisoning. Refrigeration and freezing are two of the best ways to prevent food from spoiling. Refrigeration helps to maintain the taste of fresh food. Other methods of food preservation, such as canning or drying, change the flavor and texture of the food more than refrigeration does.

Long ago, snow and ice, cool streams, springs, caves, and cellars were used to refrigerate food. The Chinese cut and stored ice as early as 1000 B.C. About 1300 A.D., Marco Polo described ices and sherbets he had eaten in the Far East. The idea of "manufacturing" ice dates back to Venice in the 16th century, when it was discovered that mixing salt and ice produced a slushy brine that remained below the freezing temperature. This could then be used to freeze clear water into solid ice. The old-fashioned ice cream maker worked on this principle.

In this old icebox (c. 1890), where was the ice placed? Can you see the drip pan?

In the United States, using ice to preserve food in homes became popular in the mid-1800s. At first, the ice trade provided "natural" ice that was cut in the winter from northern lakes and rivers and shipped to cities in the south. Eventually, wooden boxes lined with tin or zinc and insulated with various materials including cork, sawdust, and seaweed were used to hold blocks of ice to "refrigerate" food in the home. These were known as iceboxes. A drip pan collected the melt water, and emptying it daily was a common kitchen chore.

A good icebox prevented the transfer of heat from the surrounding room to the food inside. It allowed for good transfer of heat from the food to the ice. One problem with iceboxes was that the ice had to be replaced every few days.

Warm winters in 1889 and 1890 resulted in severe shortages of natural ice in the United States. This increased the use of mechanical refrigeration for the dairy and meat packing industries and for freezing and storing fish. Mechanical refrigeration was used in railroad cars and in grocery store coolers. Refrigeration also came to be used in various ways in the textile, paper, drug, soap, liquid gas, sugar, and munitions industries.

Mechanical refrigerators for the home began to appear on the U.S. market between 1910 and 1915. By 1930, half of the homes in the United States had refrigerators. This number increased to 85% by 1944. It's hard to imagine life without one, isn't it?

QUESTIONS

1. Draw a picture of a room with an icebox containing a block of ice. Think of the room, the icebox, the ice, and the food as a system. The icebox is a smaller system within the larger system of the room. On your diagram, use lines and arrows to show all of the heat transfers possible between parts of the large system. Use a solid line and arrow to represent transfers that help to keep the food cold. Use a dotted line and arrow for all other energy transfers in the system. Keeping in mind that the purpose of the icebox was to keep food cold, indicate for each solid line and arrow how the heat transfer it shows contributed to that purpose.

2. In order to make the ice in an icebox last longer, some people would wrap a blanket or newspapers around the ice to keep it from melting. Keeping in mind that the purpose of the icebox is cooling food, explain why this was or was not a good idea. Draw a diagram like the one you drew for Question 1 to explain your answer.

Workers cut ice blocks from frozen lakes and rivers in the late 1800s.

Mixing Hot and Cold Water

In Activity 2 you learned about heat transfer from a warm object, your hand, to a cold object, ice. Now you will learn one way to measure the amount of heat energy transferred.

Measure the amount of energy lost by hot water and the amount of energy gained by cool water. Compare the amounts.

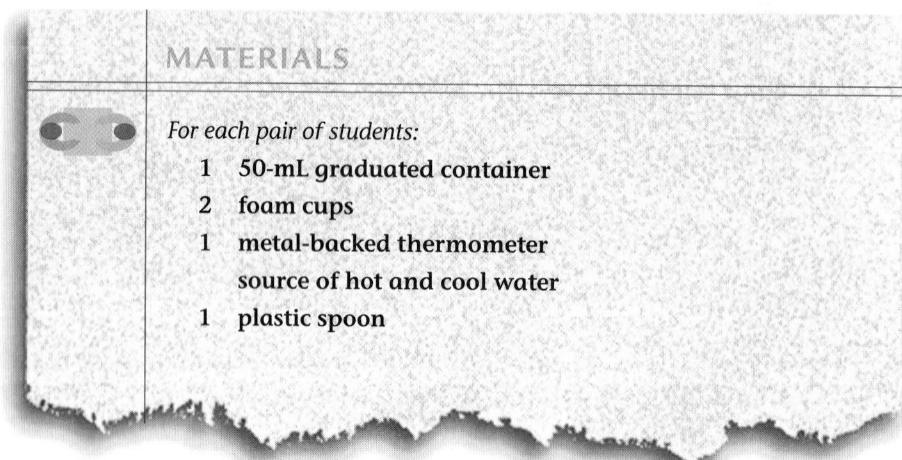

MATERIALS

For each pair of students:

1 50-mL graduated container
2 foam cups
1 metal-backed thermometer
source of hot and cool water
1 plastic spoon

PROCEDURE

1. Before conducting the investigation, describe in your science notebook what you think will happen to the temperature of the water when you mix the hot and cool water. Then prepare a data table in your science notebook, like the one shown here, to organize your observations of what actually happens.

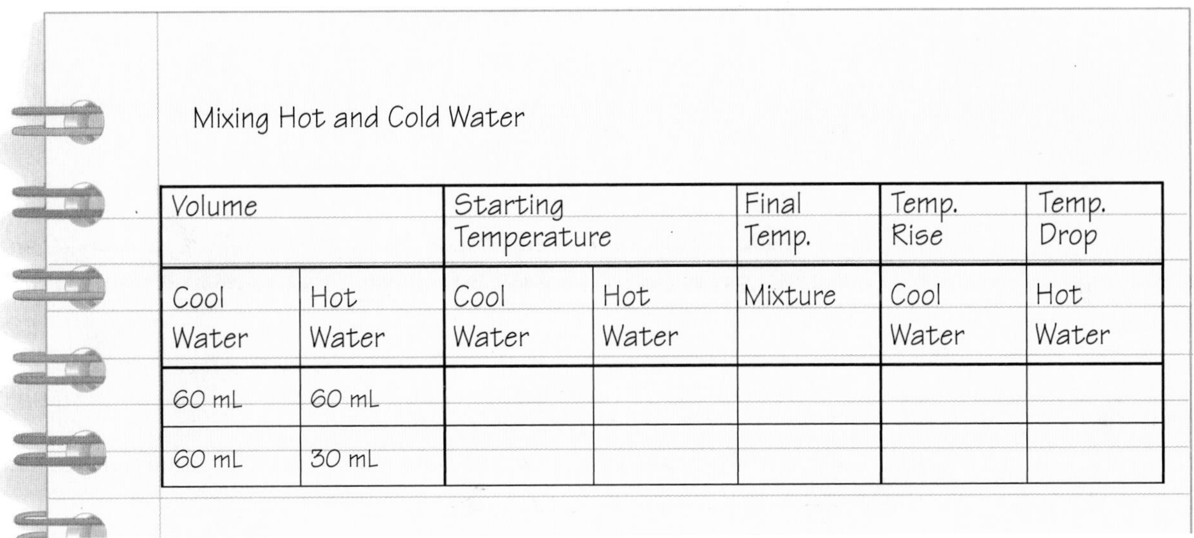

Mixing Hot and Cold Water

Volume		Starting Temperature		Final Temp.	Temp. Rise	Temp. Drop
Cool Water	Hot Water	Cool Water	Hot Water	Mixture	Cool Water	Hot Water
60 mL	60 mL					
60 mL	30 mL					

2. Follow your teacher's instructions for obtaining 60 mL of cool water and 60 mL of hot water in each of your foam cups.

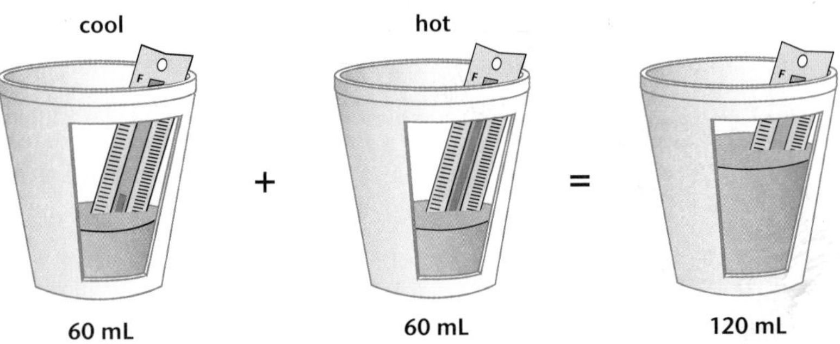

3. Measure the temperature of the cool water. Wait for a few seconds until the temperature remains steady and then record it in the data table.

4. Measure the temperature of the hot water and record it. Quickly add the cool water to the hot water. Stir gently with a spoon until the temperature of the mixture remains steady. Record the temperature in your data table.

5. Calculate the temperature change for both the hot water and the cool water. Fill in the appropriate columns in the table.

6. Now do a similar experiment but reduce the amount of hot water in half, to 30 mL. Describe in your science notebook what you think will happen before testing it. The results from the previous experiment should help you increase your ability to predict the outcome.

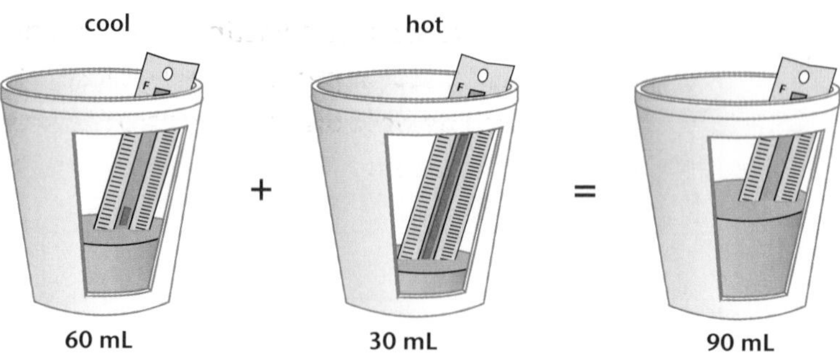

ANALYSIS

1. When you mixed equal volumes of hot and cool water, what happened to the temperature? How did the temperature rise of the cool water and the temperature drop of the hot water compare?

2. Was the result what you expected? What does this tell you about the energy transfer in this activity?

3. When you mixed only 30 mL of hot water with 60 mL of cool water, how did the temperature rise of the cool water and temperature drop of the hot water compare? How would you explain your results?

Measuring the Amount of Heat Energy Used to Melt Ice

Sometimes heat energy can be converted to another form of energy. When hot water is used to melt an ice cube, some of the heat energy of the water is transferred to the ice and converted to the internal energy of the ice particles. When enough energy has been transferred to the ice, it melts. You can figure out the amount of energy it takes to melt the ice by measuring the temperature change of the water and calculating the heat energy the water has lost.

Note: In scientific writing you will sometimes see the term **thermal energy** used in place of heat energy. Thermal means of, relating to, or caused by heat and comes from the same root word as thermostat and thermometer.

CHALLENGE

Measure the amount of heat energy needed to melt an ice cube.

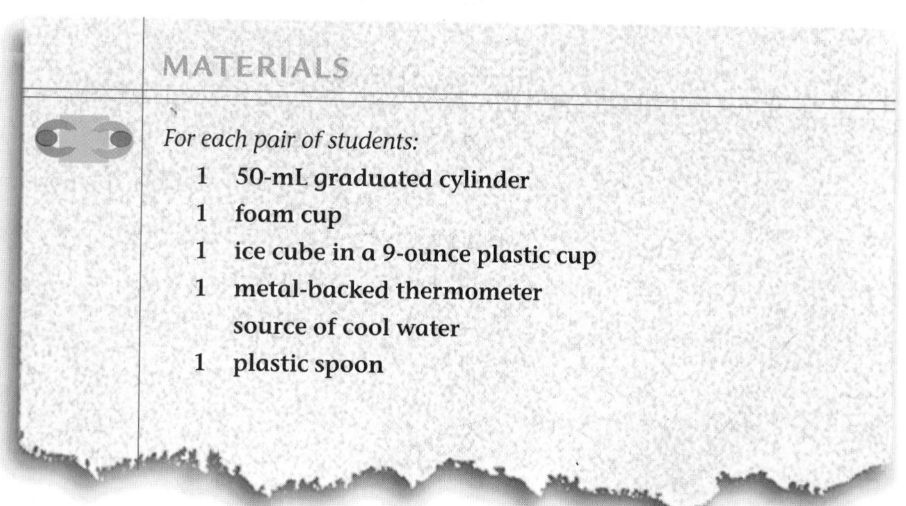

MATERIALS

For each pair of students:

1 **50-mL graduated cylinder**
1 **foam cup**
1 **ice cube in a 9-ounce plastic cup**
1 **metal-backed thermometer**
source of cool water
1 **plastic spoon**

PROCEDURE

1. Follow your teacher's directions for obtaining an ice cube and some hot water.
2. Measure 100 mL of hot water into the foam cup.
3. Record the temperature of the hot water. Then immediately shake any water off of the ice cube and place the ice in the hot water.
4. Gently stir until the ice is completely melted. (This can be difficult to see, so look carefully.) As soon as the ice has melted, record the temperature of the water.

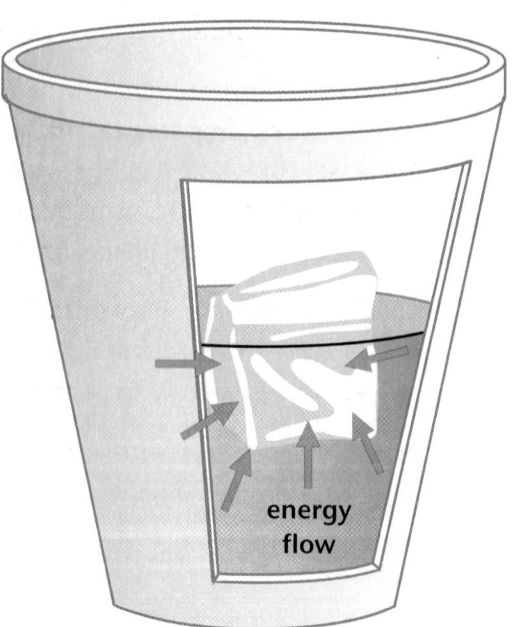

ANALYSIS

Remember, the energy required to melt the ice cube came from the heat energy of the water surrounding it. The amount of heat energy lost by the water is equal to:

mass of water (grams) × temperature change of water (°C)

1. What was the temperature change of the water?
2. How many grams of hot water were cooled? (**Hint:** You used 100 mL of hot water and 1 mL of water weighs 1 gram.)
3. Calculate the heat gained by the ice as it melted by multiplying the temperature change of the water (your answer for Question 1) times the mass of the water (your answer to Question 2).
4. Your measurement of the heat energy needed to melt the ice is based on the assumption that the energy gained by the ice is exactly equal to the energy lost by the water. Do you think all of the energy lost by the hot water was transferred to the ice? Explain. You may use a diagram similar to the one on page C-24 as part of your explanation.
5. Based on your answer to Question 4, do you think you are more likely to have overestimated or underestimated the amount of heat energy needed to melt an ice cube?
6. What effect did stirring have on the melting of the ice? What would happen if you did the same investigation again but didn't stir the ice?

Measuring Peanut Calories

A peanut contains stored energy. That energy is released when we eat and metabolize the peanut. The amount of energy we can obtain by eating the peanut is usually measured in **food Calories**. A food Calorie (with a capital "C") is equal to 1,000 calories.

CHALLENGE

Determine the number of calories in one peanut. Then convert the calorie measurement into food Calories.

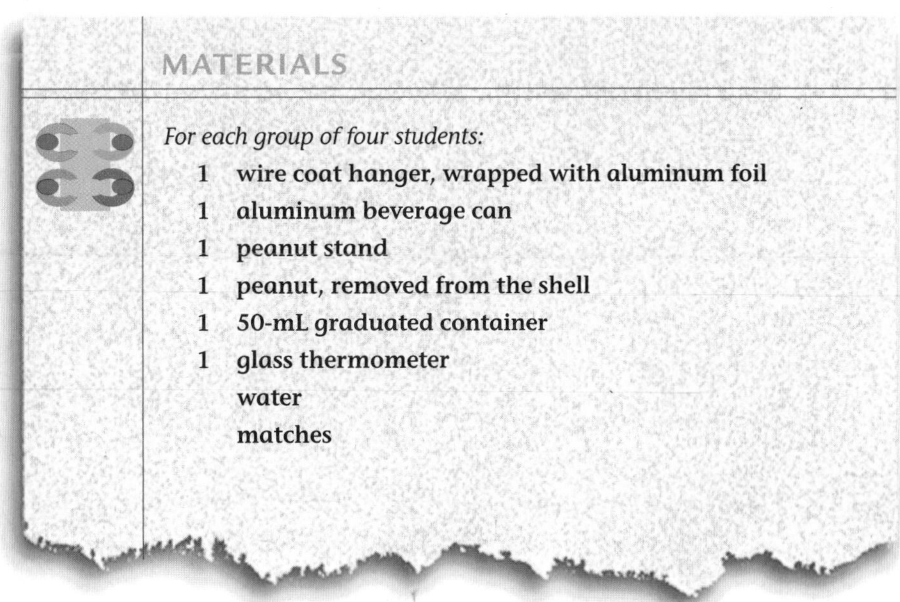

MATERIALS

For each group of four students:

1 wire coat hanger, wrapped with aluminum foil
1 aluminum beverage can
1 peanut stand
1 peanut, removed from the shell
1 50-mL graduated container
1 glass thermometer
water
matches

SAFETY NOTE: Be sure to use safety eyewear during this investigation. Have a cup of water available in case you have to put out the peanut flame. The can may become quite hot. Carefully follow all instructions from your teacher. ***Be especially careful not to get clothing or long hair near the flame.***

PROCEDURE

1. Carefully place the peanut on the peanut stand.
2. Set up your calorimeter with 100 mL water in the can.
3. Just before burning the peanut, record the starting temperature of the water.
4. Light the peanut. Once it begins to burn, slide it under the can and let it burn completely.
5. As soon as it stops burning, stir the water and record its final temperature.

The SEPUP calorimeter. Note placement of thermometer and use of foil to reduce heat loss.

ANALYSIS

Hint: If you burn a peanut, it will give off energy that you can detect as heat and light. The heat energy transferred can be measured by determining how much it raises the temperature of a known volume of water. Remember, the heat energy absorbed by water is equal to the mass of the water (in grams) times the temperature change of the water (in °C).

Before calculating the calories of heat energy given off by the peanut, you need to know (and record in your science notebook):

1. The amount of water used = _______ mL = _______ grams

   Hint: One mL of water weighs one gram.

2. The temperature change of the water = _______ °C.

3. Now you can calculate the calories by multiplying your answers for Items 1 and 2 above:

   calories = grams water × temperature change of water

   Show your calculation in your science notebook.

4. Your calorie determination is only as good as the calorimeter you used. Explain why.

5. Based on what you have learned about measuring the transfer of heat energy, how would you improve the design of the calorimeter so that you could measure all of the heat energy released by a burning peanut? Draw a detailed diagram of your improved calorimeter, and explain why it would be better than the old one at measuring the energy released by the burning peanut. Be sure to design a calorimeter that you could build yourself, if you had the materials. Label the parts of your calorimeter on the diagram.

6. Do you think your calculated peanut calorie figure is likely to be lower or higher than the actual value? Explain.

7. A food Calorie (notice the capital "C") is equal to 1,000 calories, or one kilocalorie. How many food Calories are there in one peanut?

GOING FURTHER

With your teacher's permission, determine the number of calories in a marshmallow or a puffed cheese snack.

Getting Food: Comparing Energy Used with Energy Received

Plants and animals gather energy in order to survive. Plants get their energy from the sun. A few kinds of animals are able to stay in one place and collect energy as it comes to them just the way plants do. For example, clams filter bits of organic matter that float by in the water. Most animals, however, do go out and search for food. This can take quite a bit of energy. For over a million years our human ancestors gathered energy in the form of food by hunting, fishing, and gathering. Even today, many people still hunt, fish, and gather, especially in poor countries. In some places, the energy sources (food) are so limited that there is not enough to go around, even when people travel miles in the search for food.

Energy for all life on earth ultimately comes from the sun.

Reflect on how people's methods of obtaining food have changed. Think about how this relates to our changing use of energy and its impact on Earth.

GETTING FOOD ENERGY EFFICIENTLY

Farming is a more recent human development than hunting and gathering. Researchers believe that agriculture started about 10,000 years ago. Farming allowed great numbers of people to stay in one place and do other things during the day besides hunt and gather food (like go to school). The job of farming, like all work, requires energy. Pick a type of farming (such as citrus farming in Florida, dairy farming in Wisconsin, or peanut farming in Georgia) and think about all the ways that energy is used to produce and deliver the product to market.

What would you say if the energy it took to gather or farm a particular food was greater than the energy contained in the food itself? The chart on the next page shows the ratio of energy input to food energy output for several foods produced in different ways. A ratio of 1.0 means that the input and output energy is the same. For example, coastal fishing has a ratio of 1.0—the input and output energy are about the same. But distant fishing requires much greater energy input—consider all of the ways that energy is used to get to the fishing grounds, catch the fish, process the fish, store the fish for the return, and then get the fish to market. Many modern methods of obtaining food, such as distant fishing and intensive agriculture, have high energy input to energy output ratios.

Food Source	Ratio of Energy Used to Energy Received	
	Energy Used	Energy Received
Hunting and gathering	0.1	1
Low-intensity corn	0.2	1
Range-fed beef	0.5	1
Intensive corn	0.5	1
Coastal fishery	1	1
Feedlot beef	10	1
Distant fishery	14	1

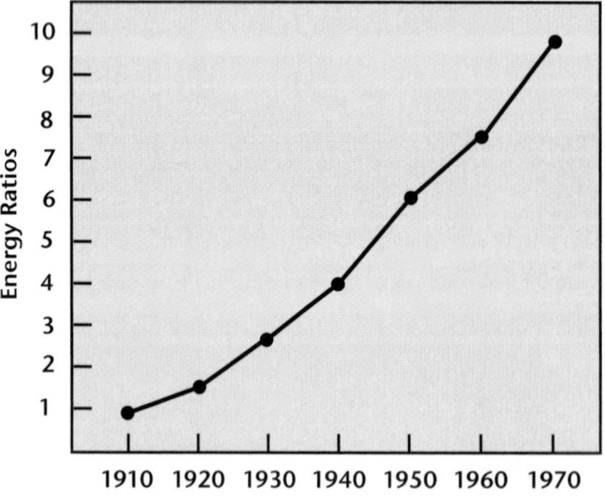

Based on data in J.S. Steinhart and C.E. Steinhart, "Energy Use in the U.S. Food System," Science, *April 19, 1974, pp. 307–315.*

QUESTIONS

1. In general, what types of food cost more energy to produce than they provide? (**Hint:** Look at the foods above the food-energy ratio of 1 to 1.)

2. Can you think of reasons why the foods above the ratio of 1 to 1 take more energy to produce than they release as chemical energy in the body?

3. With the passage of time, distant fishing has become more common than coastal fishing. Why? Why is the energy input so much greater for the distant fishing?

4 Batteries: Energy and Disposal

Batteries: Portable Energy Converters

Think of a battery as an energy converter. It is able to convert chemical energy into electrical energy. Look at the accompanying diagram of a battery and read about the various parts and how they work.

Gain a better understanding of how a battery works.

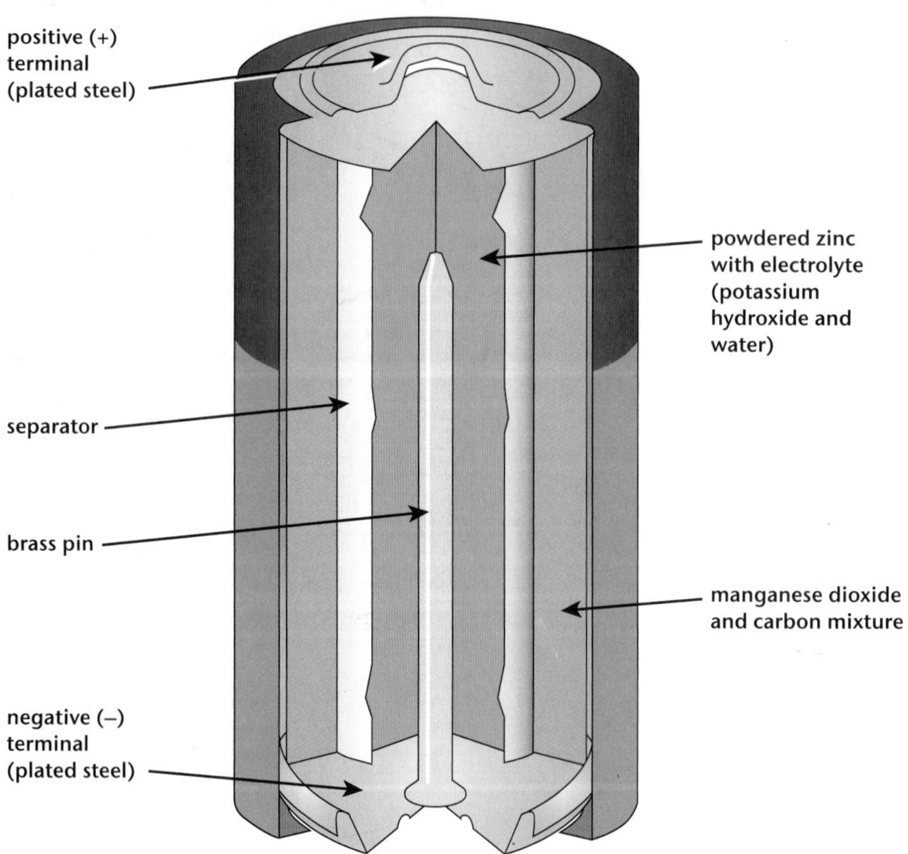

Cutaway view of an alkaline battery

ALL ABOUT BATTERIES

A common **alkaline** (AL-kah-line) flashlight battery, also called an electric cell, is pictured on the opposite page. All batteries contain a liquid, paste, or solid material that electricity can flow through. This material is called an electrolyte. The electrolyte in the D-cell alkaline battery is a moist paste of potassium hydroxide. The cell is sealed to prevent moisture from evaporating.

Batteries also have two terminals of plated steel, a positive terminal called an **anode** and a negative terminal called a **cathode**. As you can see from the picture, the positive terminal in alkaline batteries is the

Batteries and the Environment	
Single Use—Not Rechargeable (Primary Cells)	**Rechargeable Batteries (Secondary Cells)**
Zinc-carbon cell batteries *Classic, Heavy Duty, Super Heavy Duty Plus* • zinc (a toxic heavy metal) • zinc chloride (toxic by ingestion)	**Lead-acid batteries** *car, boat, motorcycle* • lead plates (toxic heavy metal) • lead dioxide plates (toxic heavy metal compound) • concentrated sulfuric acid
Alkaline batteries *Energizer®, Duracell®, Supralife®* • potassium, hydroxide solution (a very strong base) • zinc metal	**Nickel-cadmium batteries** *rechargeable, Enercell®* • nickel oxide (toxic heavy metal oxide • cadmium metal (toxic heavy metal) • potassium hydroxide solution
Lithium batteries *Ultralife®, Lithion®, Procell®* • lithium metal (causes fires) • iron dissulfide (corrodes metals, produces toxic hydrogen sulfide gas)	**Lithium-ion batteries** *cell phones, computers* **Nickel-metal hydrides** *computers, cell phones, personal digital assistants*
Mercury batteries *hearing aids* • zinc metal • mercury oxide (accumulative toxin) • potassium hydroxide	
Silver oxide batteries *watch, calculator* • silver oxide (toxic heavy metal)	

zinc anode in the center. The negative terminal is the manganese dioxide cathode connected to the outside shell of the battery.

When you put a battery in a flashlight and turn the light on, you are producing a path for electron flow that is known as an external circuit. A chemical change takes place inside the battery. The manganese dioxide cathode (negative terminal) starts to react with the moist paste, leaving an excess of electrons on this terminal. The electrons flow through the external circuit back to the zinc anode (positive terminal), which accepts them. This flow of electrons is the electrical current that passes through the flashlight bulb and produces the light.

In an alkaline battery, this chemical change cannot be reversed. When the **reactants** are used up, the battery dies. (Reactants are chemicals in the battery that react to produce free electrons.) This type of battery is called a primary cell. Other types of batteries, called secondary cells, can be recharged. The lead-acid batteries in motorcycles and automobiles are examples of secondary cells.

QUESTION

Explain in your own words why a battery "dies."

Chemical Batteries

Batteries contain materials in which chemical changes take place, converting chemical energy into electrical energy. You will construct some simple chemical batteries and explore how different metals that take part in a chemical reaction produce electrical energy. You will use a small motor to detect how much energy is produced.

CHALLENGE

Test different combinations of metals to see which combinations produce the most energy. You will also investigate what happens to the energy produced as the two metals are brought closer together. If time permits, explore how changing the concentration of the reactants affects the energy produced.

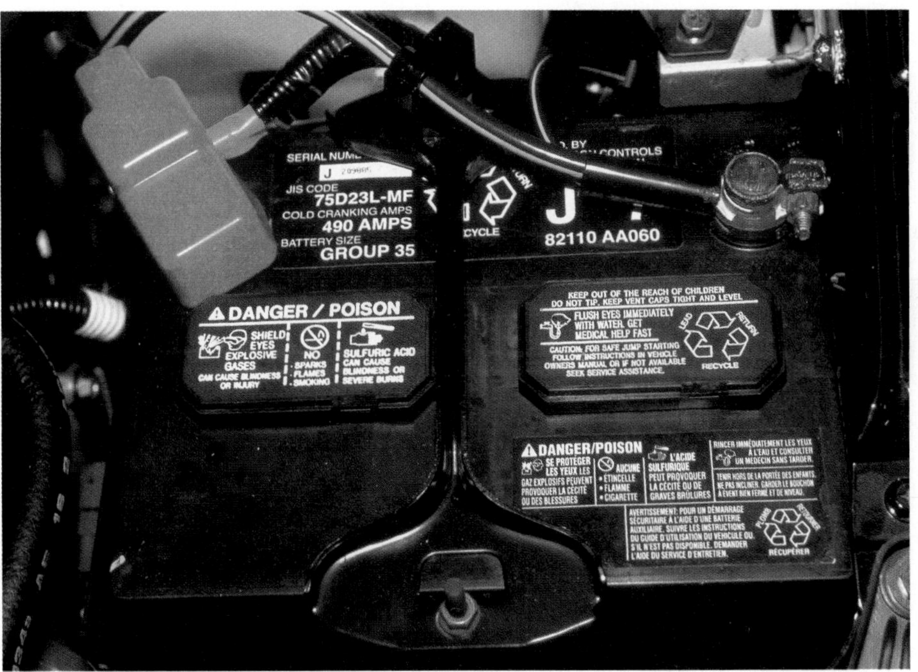

An automobile battery

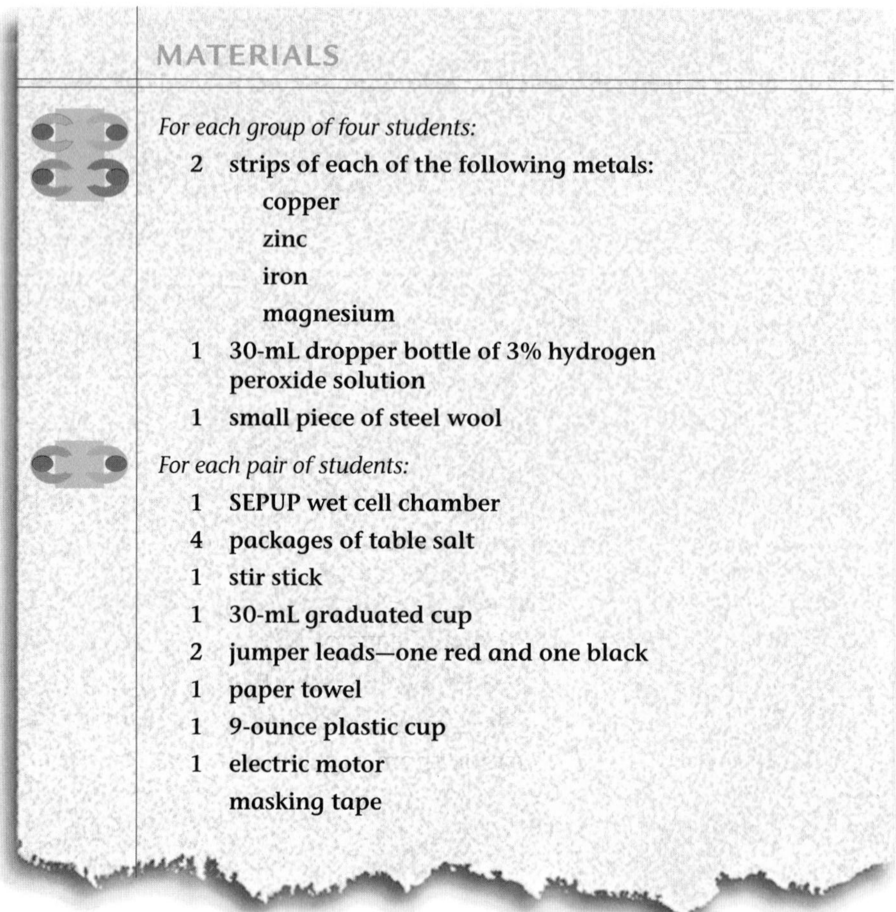

MATERIALS

For each group of four students:

2 strips of each of the following metals:
- copper
- zinc
- iron
- magnesium

1 30-mL dropper bottle of 3% hydrogen peroxide solution

1 small piece of steel wool

For each pair of students:

1 SEPUP wet cell chamber

4 packages of table salt

1 stir stick

1 30-mL graduated cup

2 jumper leads—one red and one black

1 paper towel

1 9-ounce plastic cup

1 electric motor

masking tape

SAFETY NOTE: Be sure to use safety eyewear during this investigation. Hydrogen peroxide can stain your clothes.

PROCEDURE

Part One: The Effects of Different Metals

The first goal of this investigation is to discover which two metals produce the most electricity when set up according to the directions on the next page. Be sure to try all possible combinations. Use the list of combinations you copied from the class discussion to guide your testing. As you test each combination, share your observations with the other members of your group and record the results in your science notebook.

Two members of your group should test the three combinations of metals with copper. The other two group members should test the other three combinations.

Note: It is extremely important to dry the metals on a paper towel and clean both sides before using them for the next test!

1. Add water to near the top of the SEPUP wet cell. Then pour the water into a plastic cup. Add 4 packages of salt. Add 25 drops of hydrogen peroxide. Stir well and pour back into the SEPUP wet cell.
2. Put a small piece of masking tape on the motor shaft.
3. Pick two of the metals. Clip a wire onto each metal and then clip the other ends to the motor.
4. Lower the pieces of metal into the slots of the SEPUP wet cell as shown. Leave a little sticking out of the cells. Observe how fast and how long the motor spins. If the motor still spins after 2 minutes, stop timing. Remove the two metal pieces. Dry them, and then shine them with a piece of steel wool.

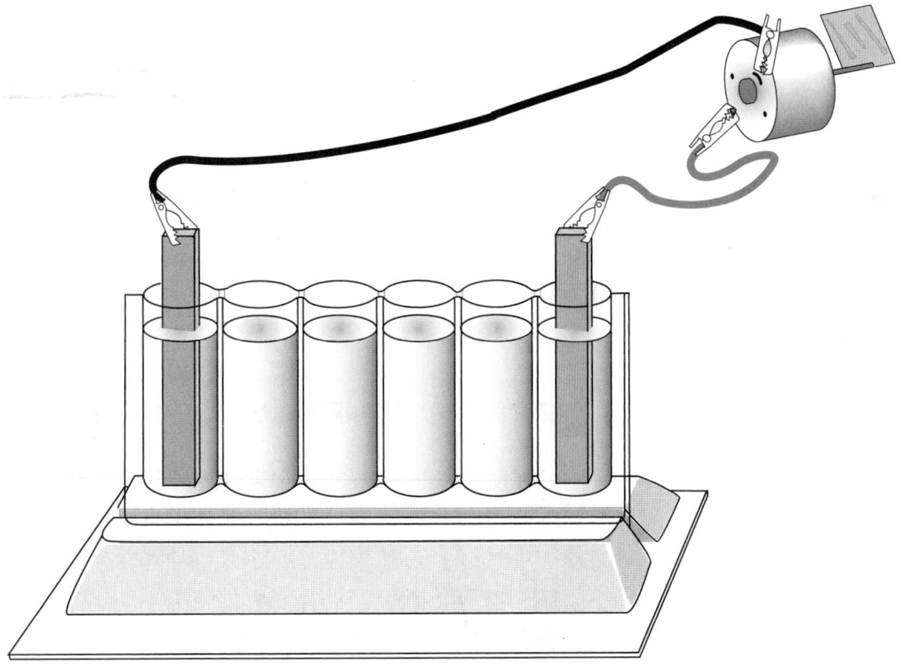

The SEPUP wet cell chamber.

Part Two: The Effect of Distance Between Metals

As a group, determine what happens to the energy produced by the battery when two metals are placed closer to each other. For this test use the zinc and iron strips as your two metals. Place them in the slots at opposite ends of the SEPUP wet cell. In steps, move the zinc strip from slot to slot so that it gets closer to the iron strip. Make a table in your science notebook to record the results.

Part Three: The Effect of Reversing Connections

Explore what happens to the direction the motor turns when you reverse the connections on one pair of metals. Use the zinc and copper combination of metals to explore this. What does this tell you about the direction (path) that electrons are flowing in the circuit?

Part Four: Effect of Metal Surface Area

Use the zinc and copper combination to explore what happens to the energy produced by the battery as you gradually remove one of the metal strips from the battery solution.

ANALYSIS

1. Construct a data table in your science notebook to record your observations.
2. From your observations, which combination of metals produced the most electrical energy? The least energy? What is your evidence?
3. How does placing the metal strips closer to each other affect the speed (the amount of energy produced) of the motor? Propose an explanation for this.
4. How does the amount of the metal exposed to the solution change the energy produced?
5. If you were given copper, silver, and nickel to make a battery, what kind of experiment could you do to find out which combination of two of these metals would produce the most energy? Explain how you came up with your design.

GOING FURTHER

If you have time, investigate one of the following questions. First, design an investigation and write up the procedure in your science notebook. After your teacher approves your design, do the investigation and record your results. Write a one- or two-paragraph summary of what you did and what you can conclude from your investigation.

1. How does changing the concentration of the salt or hydrogen peroxide affect the amount of energy produced?
2. What other combinations of metals can be used to power the motor?

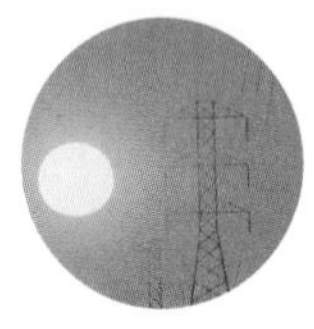

You'll Get a Charge Out of These!

You already know that batteries and electricity are related to each other. The reading below will tell you about how batteries were discovered and about how many batteries you would need to provide electricity for different events.

What are batteries? Who first discovered them? How many batteries would it take to keep your heart beating? Read on to find answers to these questions.

THE FIRST BATTERIES

Life without a portable CD player, electronic game, or a flashlight may be hard to imagine. Yet, a little less than 200 years ago, the device that powers each of these was unknown.

Eighteenth-century scientists like Benjamin Franklin and Luigi Galvani provided early clues to the possibility that people might be able to make use of electricity. For example, Franklin tied a key to a kite string to prove that lightning was actually electricity. The experiments of Italian scientist Luigi Galvani led to ideas that helped develop modern batteries. Galvani was a professor of anatomy at the University of Bologna in 1786. While dissecting a frog, he noticed that when the frog's muscles were touched with two strips of different metals, zinc and copper, the legs twitched—they actually moved! He also noticed that the sparks from a nearby electric machine made the frog's legs twitch. Galvani explored this further and called the cause of this movement "animal electricity." Although this idea of animal electricity was discredited, Galvani's work led to the discovery of electrical currents, whose action is described by the verb "galvanize."

Later, in 1800, Count Alessandro Volta of Italy invented the battery—the first source of electricity that could produce a current. This invention made the development of the telegraph and telephone possible.

The chart below will give you an idea of how electrical power measured in kilowatt-hours is related to some common events. It will also tell you how many D-cell batteries would be needed to produce the same amount of energy.

Battery Power

Kilowatt-hours	Energy Activity	Equivalent No. of D-cell Batteries
1,000,000	Energy produced by a large electrical generating plant per hour	50,000,000
100,000	Energy consumed per person in the U.S. each year	5,000,000
10,000	Energy needed to refine one ton of steel	500,000
10	Energy consumed by a car trip of 40 kilometers (24 miles)	500
1.0	Average person's daily intake of food energy	50
0.1	Energy used by an average person's physical work in one day	5
0.01	Energy produced per hour by a flashlight	0.5
0.00001	Energy produced per hour by the human heart	0.0005

Which Battery to Choose?

Each year in the United States, close to 2 billion batteries are thrown away. Think about that number—2 billion. It's a very large number, and it represents a lot of waste, yet to reach it more than 272 million people in the U.S. would each have to throw away just seven or eight batteries a year.

CHALLENGE

Decide how you would you select a battery to use. Compare the costs, both economic and environmental, to help answer this question.

Discarded batteries often contain harmful chemicals.

BATTERY DISPOSAL

To give you an idea of why so many used batteries are thrown away, let's consider a simple example. Imagine one D-cell battery can be used to run your portable radio (a large radio requires more like 4 to 8 batteries). Look at the chart below to see how many changes of D-cell batteries would be needed to run your radio for 500 hours. Consider the costs. See Question 1 on page C-46.

Battery Type	Cost per Battery	Hours Each Battery Lasts	No. Needed for 500 Hours Usage	Total Cost
Alkaline	$1.30	20.0	25	$26.00
Rechargeable	$3.40	2.5	1	$20.62*

*includes charger unit and cost of electricity to recharge

Was your decision based on cost? Was it based on convenience—how many times would you need to buy batteries or replace them, or how long your radio would run before you would have to replace the battery? What about batteries as wastes? How might they affect the environment?

Many batteries end up in community landfills and incinerators where they can be a hazard to the environment. Until recently, most non-rechargeable batteries contained mercury to help the cell perform well over long periods of time. Rechargeable batteries, while they do not contain mercury, do contain nickel and cadmium. All of these metals are classified as toxic heavy metals.

In 1985, discarded batteries added 46,553 pounds of mercury to our nation's landfills, according to the United States Environmental Protection Agency (EPA). Well into the 1990s, after the toxic nature of mercury was better understood, mercury use in battery production fell sharply, from 21,410 pounds in 1990 to 1,500 pounds today. However, as mercury use has dropped, the use of cadmium has risen. In 1985, only 796 pounds of cadmium were used in batteries, while today over 13,800 pounds are needed. Why? Cadmium is an important ingredient in rechargeable nickel-cadmium batteries, present in laptop computers, cameras, and many portable electronic devices. So, a trade-off in the use of rechargeable batteries is to use less mercury but more cadmium.

While we need to do all we can to reduce the overall amount of toxic heavy metals like mercury and cadmium in landfills, it is important to compare amounts and sources. In 1993, more than 2 million pounds of cadmium were released to the atmosphere in the U.S., mostly from zinc and copper smelting (smelting is the breakdown of metal ores using heat). And most mercury in the atmosphere today comes from coal-fired power plants.

In response to laws, the urging of environmental groups, and their own concern about the environment, battery manufacturers have begun an aggressive effort to collect and recycle nickel-cadmium batteries. Similar programs were begun earlier to collect mercury batteries used in hearing aids. Some states, like New Jersey, have passed tough battery recycling laws and have required collection centers for all types of batteries.

QUESTIONS

1. Think carefully about which battery you would use in the radio described in the first paragraph on the previous page. What are your reasons for selecting it? Write these in your science notebook.
2. What kind of plan does your community have for recycling used batteries?
3. Earlier you read that about 2 billion batteries are discarded each year. Because all of these batteries contain small amounts of heavy (toxic) metal, they increase the risks of contaminating the environment and exposing people to a greater chance of getting sick from the metals.
    - **a.** What kinds of information would you need to assess the risk of disease from heavy metals in batteries?
    - **b.** Assuming that the risk is too high, what are some of the ways you could reduce the risk? List as many ways of reducing the risk as you can.
    - **c.** Now that you have a list of possible methods of reducing the risk, you need to make a decision about what action(s) you will take. What are some of the issues that you need to consider in order to decide what to do to reduce the risk?

Appliances in 1902

You will conduct a survey of the appliances that you, your parents, and your grandparents have used over the generations. Before beginning the survey, look at the illustrations of appliances from the Sears, Roebuck and Co. catalog of 1902 shown below and on the next page.

Try to identify all of the objects from the Sears, Roebuck and Co. catalog. What kinds of energy were required in the home in 1902? What do you think was the main source of home energy in 1902?

CLEAN SHAVE

Charcoal Irons.

No. 23R408 Family Charcoal Irons, with removable top, wood handle with tin shield, one flue. Weight, 6¼ pounds.
Price, each..........................85c

Visible Writing Machine.

$22.50 buys this magnificent thoroughly up to date Visible Typewriter.

SUN

THE KODAK.

ANYBODY can use the KODAK. The operation of making a picture consists simply of pressing a button. One hundred instantaneous pictures are made without reloading. No dark room or chemicals are necessary. A division of labor is offered, whereby all the work of finishing the pictures is done at the factory, where the camera can be sent to be reloaded. The operator need not learn anything about photography. He can "*press the button*"—*we do the rest.*

PRICE, $25.00.

Send for copy of KODAK Primer, with sample photograph.

The Eastman Dry Plate and Film Co.,
ROCHESTER, N. Y.

The Acme Combination Washer.

This Machine is Warranted not to Leak.

In Use $4.25 Open

The 20th Century Gas Headlight.

Burns any loose carbide six to eight hours with one charge. aluminum parabola reflector, which spreads a wonderful light. A brass carbide holder accompanies every lamp, together with full directions for handling same. New adjustable lamp bracket, to fit either head or fork.

No. 19R359 20th Century Gas Headlight.
Price..(If by mail, postage extra, 38 cents)..$2.25

Calcium Carbide.

For use in Acetylene Gas Bicycle Lamps. ¼-inch size, packed in airtight tin cans containing 2 pounds.

No. 19R364 Carbide, 2-lb. can.
Price, per can........................22c
Not mailable.

The Cem Magic Lantern.

Acme Wonder Wood Cook.
WITH PORCELAIN LINED RESERVOIR.

THE IRON ACE
FOR COAL ONLY.

MADE IN SEVEN SIZES, from $2.70 to $11.95, according to size, delivered on the cars at our western foundry on the Upper Mississippi River.

Has large ash pit, cast foot rails, annular draft ring admitting the draft equally all around the shaking grate, opening and closing automatically as you shake the grate. Th[illegible] is so constructed that an extension sheet iron drum can be attached at any time.

Cash with order prices, delivered on the cars at our Mississippi Valley foundry.

$10.45

$10.45

Our Overhead One-Horse Power.

This style of horse power is very convenient and popular, because, owing to its construction, it has many advantages not found in down powers. It is especially adapted for use in a barn where several horses are kept, or in small livery stables. The power can be bolted to the timbers above the driveway and machines can be set on the floor either above or below the power. When not in use the center post can be lifted from its socket and put out of the way, leaving the floor clear for other purposes. Then when power is to be used again all that is necessary is to set the post in place, hitch the horse to the sweep and go ahead. The center post which we furnish is made of 6-inch by 6-inch timber and is 12 feet long. It is amply strong and can be cut to any desired length. The 1¼-inch driving shaft, to which the pulley is attached, is regularly made so that the measurement from center of master wheel to center of pulley face is 3 feet 3 inches, but additional shafting can be coupled to this shaft so as to change position of pulley or allow the use of other pulleys. The driving pulley is 18 inches in diameter with 3-inch face and makes 37½ revolutions to one round of the horse, or about 135 revolutions per minute. Additional shafting or change in size of pulley is extra. Length of sweep from center of post to eye-bolt is 7 feet 6 inches. For driving small feed cutters, corn shellers, feed grinders, wood saws, etc., this power cannot be excelled. Weight, 450 pounds. Shipped direct from factory in Southeastern Wisconsin.

No. 32R1823 Overhead Horse Power. Price.....................$18.45
No. 32R1824 Extra Line Shaft, per foot..........................32

Our Complete Learner's Telegraph Outfit.

PRICE $1.65

Utility Ads Then and Now

You will read two ads about energy: one of them appeared over 30 years ago and the other ad appeared in 2002. Study them to learn how advertisers and people's attitudes about energy have changed.

CHALLENGE

Find out how people thought about energy at the time each of the ads was written.

Need a little help with the housework?

All of us need a little help around the house now and then. But chances are you'll need less if you make regular use of dependable electric appliances.

Whether you need help with the cooking, washing, cleaning—you'll find a way to do the job quickly, safely, and easily—and for just pennies per day!

Spend less time with the housework—and more time with your family. Put electricity to work—for you and for your life. You'll be glad you did.

ELECTRIC LIVING IS BETTER LIVING

SAVING ENERGY MAKES GOOD CENTS

While compact fluorescent bulbs can cost more initially, they save you more money in the long run—and save more energy in the process.

What's more, there are many other steps you can take to reduce your energy costs:

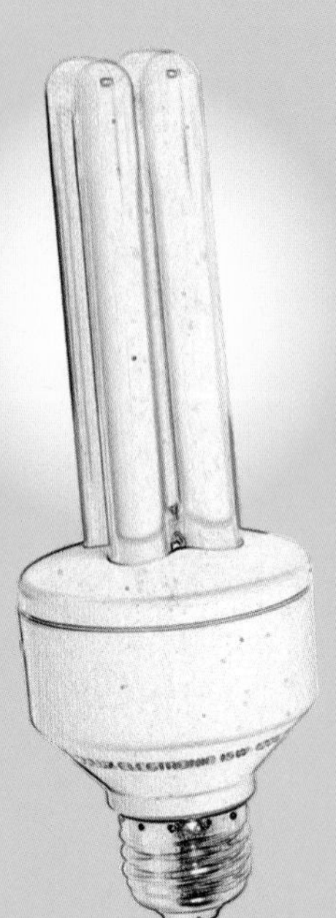

- Lower your thermostat to 68–72 degrees, year-round.
- Insulate your hot-water heater and pipes.
- Install low-flow shower heads to reduce hot water consumption.
- Add extra insulation in your attic or ceiling crawl space.
- Turn off all unnecessary lights when you leave the house.
- Replace old weatherstripping around doors and windows.

These changes can cost you pennies—and you'll save big over time. To find out more ways to make your home more energy efficient, call our customer service office.

SILVER OAKS GAS AND ELECTRIC

QUESTIONS

1. What is the message in the old ad? Give at least two specific examples of this message.
2. What is the message in the 2002 ad? Give one example.

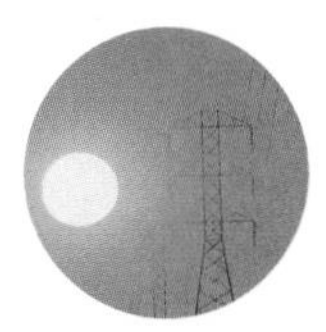

Then and Now Survey

As you conduct this survey, you will collect data about the number of appliances you have in your home today, compared with the number of appliances your parents or guardians had at home when they were your age. The data will tell you about changes in the use of appliances and energy.

CHALLENGE

Take a look at the appliances listed on the survey distributed by your teacher. Guess how many you have in your home right now. Now conduct the survey and find out.

PROCEDURE

1. In Column A, write the number (how many) of each kind of appliance you have in your home today. If you have an appliance that is not on the list, write its name on one of the blank lines.
2. Next, have a parent, or an adult you live with, fill in Column B with the number of each kind of appliance in the home when he or she was your age. If you can, ask a grandparent, an older neighbor, or other senior citizen to fill in Column C for the number of each kind of appliance in the home when he or she was your age.
3. When you have numbers filled in for Columns A, B, and C, add the numbers in each column to get the total number of appliances in the column for each category. Write the totals in each of the boxes.
4. Calculate a grand total by adding together the totals in each category. Fill in the grand totals.

ANALYSIS

1. What is the total number of appliances that you have in your home today? How many appliances did your parent or guardian have when he or she was your age?

2. If your grandparent or an older neighbor was available to survey, how does the number of appliances in your house today compare with the number reported from this third generation?

3. What might be some reasons for the changes in the number of appliances over two or three generations? Which category of your survey had the largest changes?

6 Developing an Energy Savings Plan

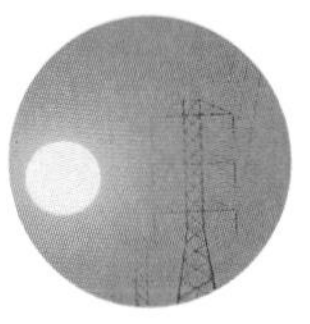

In this activity you will take another look at your top ten appliances lists. You will determine how much electricity each appliance uses in a year and calculate the total cost of the electricity.

A residential electric meter.

CHALLENGE

Come up with a plan to save 20% of the cost of electricity on one of the appliances on your top ten lists.

PROCEDURE

1. Locate the two top ten lists that you recorded in your science notebook for Activity 5, "Electric Appliances: Then and Now Survey."

2. Add four columns with these titles to each list: Power (watts), Time Used per Year, kwh of Electricity, and Cost of Electricity

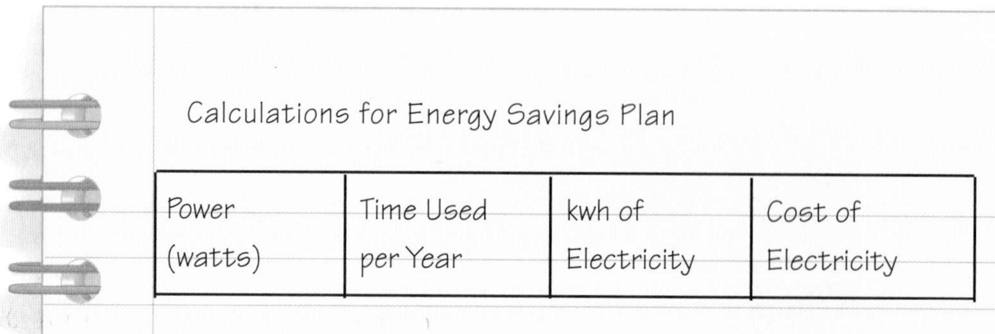

Power (watts)	Time Used per Year	kwh of Electricity	Cost of Electricity

3. Follow your teacher's directions about how to get the information needed for each column. You need to determine the power, time used per year, the kwh of electricity used, and the cost of the electricity for each appliance on your lists, using either Method A or Method B.

Method A (Using Your Own Numbers)

a. Sometimes you can find the power rating (in watts) by looking for a metal tag or label on the back or bottom of the appliance. Copy the number in watts that appears there.

b. You must estimate the amount of hours this appliance is on in a year. (**Hint:** Start with how many hours you think it is on in one day, and multiply by 365 to convert to a year. Alternatively, take the number of hours the appliance is on in one week, and multiply it by 52 to convert it to a year.)

c. Next find the kwh, or kilowatt-hours, of electricity for each appliance by using the equation:

(watts/1,000) × time = kwh of electricity

d. Find the cost for each amount of electricity. Multiply the kwh by the local cost of electricity.

Method B (Using the Table)

a. Refer to the tables on pages C-55–57. Find each appliance from your top ten list. If an appliance is not listed, try to find something on the list that is similar to your appliance. If that is not possible, try to get the information using Method A.

b. Copy the amounts—Power (watts), Time Used per Year, and kwh per Year—that appear for each appliance into the proper columns in your science notebook.

c. Find the cost of the amount of electricity that each appliance uses. Multiply the kwh by the local cost for electricity where you live.

d. Now, add together the cost of electricity for each appliance to determine the total cost of electricity used by all of the appliances on each of your lists. Write this number at the bottom of the lists.

Kitchen			
Appliance	**Power (watts)**	**Hours Used per Year**	**kwh per Year**
Blender	390	40	15.6
Coffee maker	900	120	108
Waffle iron	1,100	20	22
Toaster	1,200	35	42
Broiler	1,400	70	98
Range	12,200	100	1,220
Microwave	1,450	130	189
Dishwasher	1,200	300	360
Refrigerator (12 cu ft)	240	3,000	720
Frostless refrigerator (12 cu ft)	320	3,813	1,220
Freezer (12 cu ft)	340	3,500	1,190
Frostless freezer (15 cu ft)	440	4,000	1,760

Laundry			
Appliance	**Power (watts)**	**Hours Used per Year**	**kwh per Year**
Iron	1,000	140	140
Washing machine	500	200	100
Clothes dryer	4,800	200	960
Water heater	2,500	1,600	4,000
Quick-recovery water heater	4,500	1,000	4,500

Comfort			
Appliance	**Power (watts)**	**Hours Used per Year**	**kwh per Year**
Electric blanket	180	828	149
Air conditioner	900	1,000	900
Dehumidifier	250	1,500	375
Humidifier	180	900	162
Fan (attic)	370	800	296
Fan (window)	200	850	170

Health and Beauty			
Appliance	**Power (watts)**	**Hours Used per Year**	**kwh per Year**
Toothbrush	7	60	0
Hair dryer	750	50	37.5
Shaver	14	80	1.12
Sun lamp	280	60	16.8

Lighting			
Appliance	**Power (watts)**	**Hours Used per Year**	**kwh per Year**
Light bulbs (in home)	660	1,515	1,000

Entertainment			
Appliance	**Power (watts)**	**Hours Used per Year**	**kwh per Year**
Radio	70	1,200	84
Computer	480	1,080	518
TV (color, solid-state)	200	2,200	440
VCR	120	333	40
Spa/hot tub	1,940	1,460	2,750
Pool pump	1,380	1,848	2,830

Housewares			
Appliance	**Power (watts)**	**Hours Used per Year**	**kwh per Year**
Clock	2	8,760	17.5
Vacuum cleaner	630	75	47
Sewing machine	75	140	10.5

ANALYSIS

1. Which appliances on your lists cost the most money to operate per year? Which cost the least amount of money?
2. Which list, personal or family, shows the greatest costs? Explain the differences between the lists.
3. Follow your teacher's directions to develop an energy savings plan to reduce 20% of the total cost of electricity used by appliances on a list. (**Hint:** You need to first calculate how much money you need to save. Then look at the individual appliances on the list and think about reducing the use of each.)
4. How much money will your energy plan save?
5. Explain the choices you made for your energy savings plan. Be sure to describe the evidence you used to make your choices as well as the trade-offs or conflicts you resolved.

7 Electrical Energy: Sources and Transmission

Where Does Electricity Come From?

There are many ways to generate electricity. Some use renewable resources such as biomass (wood and other plant materials), while others use nonrenewable resources such as coal, natural gas, and fuel oils.

Suggest ways to generate electricity and explore some advantages and disadvantages for each energy source.

U.S. Electrical Energy Sources in 2001

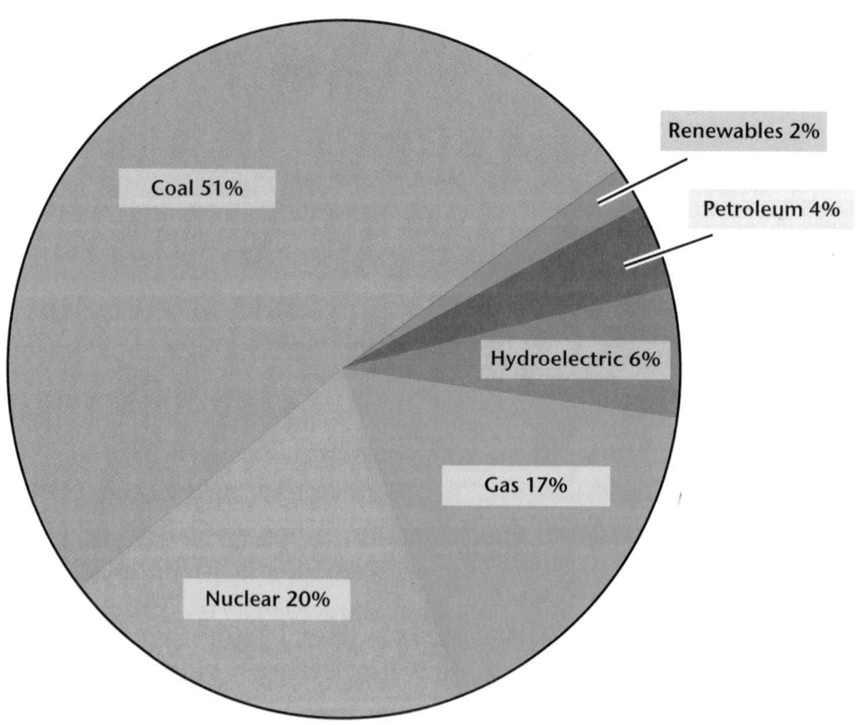

Source: U.S. Department of Energy (November 2001)

Many energy sources to can be transformed into electricity. This is usually, but not always, done by burning or otherwise creating heat to produce steam to drive an electrical generator.

Renewable Energy Sources	
Solar thermal (T) energy	Mirrors and reflectors concentrate the sun's heat.
Solar photovoltaic (P) energy	Solar cells (batteries) convert the sun's energy directly.
Wind energy	Wind provides kinetic energy.
Hydroelectric	Flowing or falling water provides kinetic energy.
Geothermal	The earth's core generates heat energy.
Biomass	Wood, paper, or municipal wastes can be burned to generate energy.
Municipal solid waste (MSW)	A city's trash can be burned to generate energy.

ELECTRICITY UPDATE

Electricity use in the United States has soared. In the 1950s, fewer than 400 million kilowatt hours supplied our electricity needs. Today, we consume more than 3.7 trillion kilowatt hours. Where does all this electricity come from? In the 1950s almost all electric power came from large government-regulated companies called utilities. The table shows that now other companies, called non-utilities, provide a growing share of our electricity.

Electricity Production, in Thousands of Kilowatt Hours			
Year	**Total**	**Utilities**	**Non-utilities**
1991	3,071,201	2,825,023	246,178
1995	3,357,837	2,994,529	363,308
2000	3,799,944	3,015,383	784,561

Wind Energy

Wind can turn the blades of large turbines to produce electricity. From 1880 to 1930, over 6 million windmills generated electric power in rural areas of the western United States. Then, a national project to bring electric lines to rural areas made most of these windmills unnecessary. Today, large assemblies of windmills, known as wind farms, have been set up in areas of the country where the wind blows for long periods of time. Because each wind turbine produces a relatively small amount of electricity, vast areas of land are needed in order to produce significant amounts of electric power.

Solar Energy

Energy from the sun can be converted directly into electricity by solar cells known as **photovoltaic** (foto-vole-TAY-ic) cells. It can also be used to heat water and other liquids, and this heat, or heat energy, can be transferred to things such as swimming pools, hot-water heaters, and turbines that produce electricity. It takes a large number of solar collectors, and therefore large areas of land, to collect enough solar energy to generate power for large-scale applications or to create a lot of heat.

Hydroelectric Power

Hydroelectric power generators use water to spin a turbine connected to an electricity generator. The water may come from rivers, streams, or lakes. Construction of a dam is often necessary. The kinetic energy in the flowing or falling water is converted to mechanical energy by the turbine and then into electricity by the generator. Some people are concerned about the environmental effects of damming streams and rivers.

Biomass Energy

Power plants that burn biomass (wood, paper, or municipal waste) to produce power are similar to power plants that burn coal, oil, and natural gas. The fuels are burned in a boiler and produce steam to drive a conventional steam turbine. Burning biomass to produce energy raises some of the same environmental concerns as incinerators.

Fossil Fuels (Coal, Fuel Oil, and Natural Gas)

Coal is the most abundant fossil fuel in the United States and the leading source of electricity. Coal is powdered into a fine dust and injected into the furnace to power a conventional steam boiler. Fuel oil and natural gas are also used as fuels to boil water. The gases produced when coal burns—sulfur dioxide and nitrogen oxides—have been the major producers of acid rain. Natural gas produces only water and carbon dioxide when it is burned "cleanly." Fossil fuels are nonrenewable energy sources.

Geothermal Energy

Geothermal energy is heat derived from the earth (the prefix "geo-" means earth). It is the heat energy contained in the rock and in fluid that fills the fractures and pores within the rock in the earth's crust. It is usually used to heat water to produce steam to drive conventional turbines.

Nuclear Power

When atoms of uranium are split apart (fission reactions), the heat generated can be used to boil water and produce steam to power a turbine. Concern over the environmental effects and accidents, such as the accidents at Chernobyl in the former Soviet Union (1986) and Three Mile Island in Pennsylvania (1979), have brought construction of new plants in the United States to a standstill. Nuclear power plants do not produce the air pollution of fossil fuel burning plants. They do generate radioactive waste which requires special facilities that must contain the waste safely for thousands of years.

QUESTION

Your city needs more electric power to provide more efficient services and so that more businesses can be built to create new jobs and. You have been asked to recommend a source of power. Considering all of the different sources, which one would you choose and why? Be sure to identify the various sources of electrical power available as well as the factors you would consider in making your choice.

Comparing Costs of Generating Electrical Energy with Environmental Costs

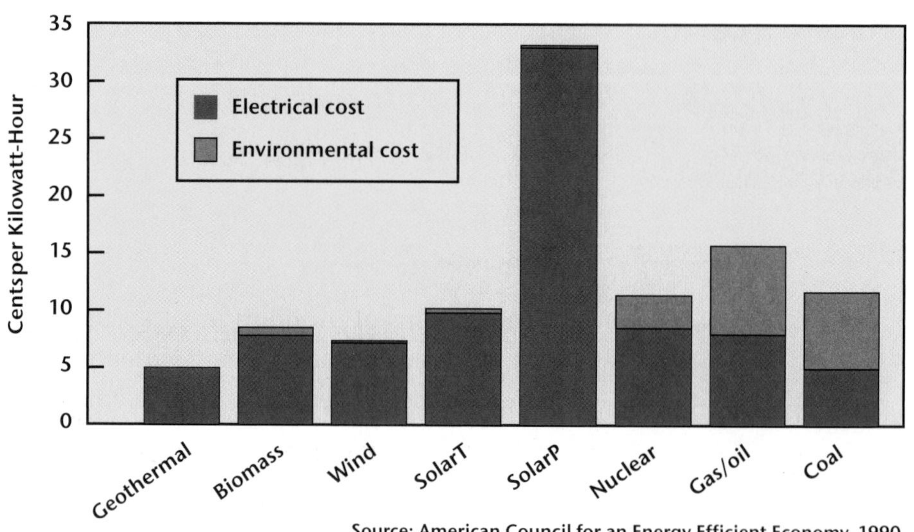

Source: American Council for an Energy Efficient Economy, 1990

Hot Bulbs

In this activity you will calculate how efficiently a flashlight bulb produces light energy. The more efficiently the bulb produces light energy, the less wasted energy it produces in the form of heat. Be sure to record and label all measurements and calculations in your science notebook. A chart similar to the one on page C-65 will help you keep organized.

How good is a flashlight bulb as a heater? Do the activity to find out!

MATERIALS

For each group of four students:

- 1 9-volt battery, with harness and connection clips
- 1 flashlight bulb with socket and white plastic cover
- 1 SEPUP tray
- 1 graduated cylinder
- 1 thermometer

SAFETY NOTE: Do not try this investigation with any other kind of battery without consulting your teacher. Never, under any circumstances, place plugged-in electrical appliances in or near water.

PROCEDURE

1. Using the graduated cylinder, carefully measure 12 mL of water into Cup A of the SEPUP tray. Measure the initial temperature of the water and record it in your science notebook.
2. Use Cup B as a control for this experiment. Decide in your group what should be placed in Cup B and what measurements should be taken for the control.
3. Insert the flashlight bulb into the white plastic cover. Be sure that the concave side of the plastic cover faces up and the bulb faces down. Screw the brass socket onto the bulb. Insert the thermometer into the plastic cover as shown.
4. Connect the 9-volt battery to the socket, using the connectors provided. Place the lighted bulb into the water in Cup A for exactly 3 minutes. Time this as precisely as you can.
5. After 3 minutes, remove the bulb from the cup. Measure the final temperature of the water and record it.

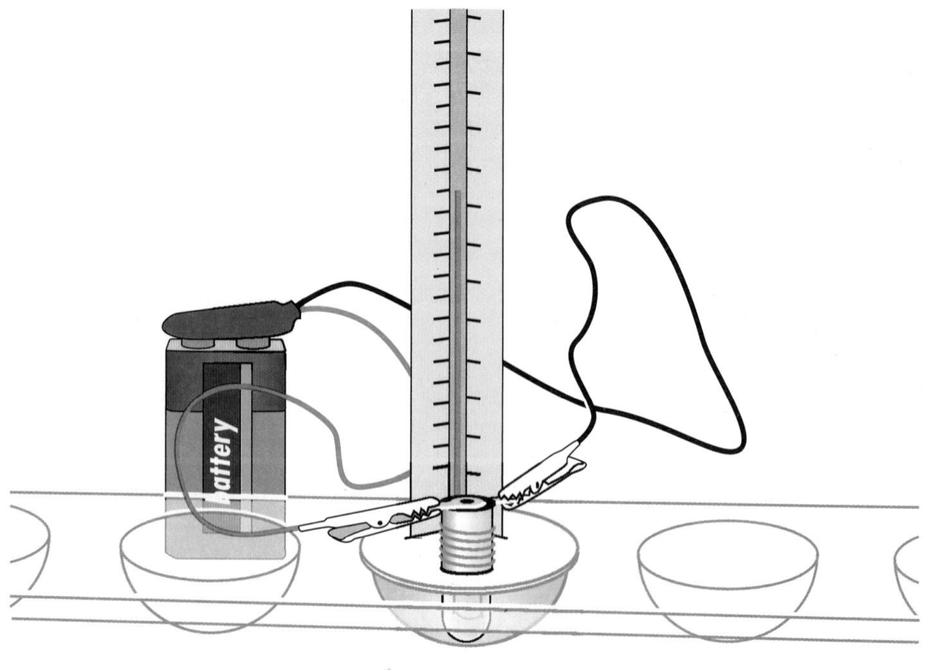

ANALYSIS

1. Calculate the temperature change of the water (final temperature minus the initial temperature).

2. Calculate the heat energy output of the flashlight bulb (in calories) using the equation:

   heat energy output (in calories) = volume of water × temperature change

3. Using the given energy input of 82 calories in 3 minutes, calculate the waste heat production of the flashlight bulb using the equation:

   % waste heat = (heat energy output / energy input) × 100

   Record and label the calculation in your science notebook.

4. Now use the heat efficiency calculation you just made to state the light bulb's light efficiency.

Heat Efficiency of a Flashlight Bulb

Measurements and Calculations	Experiment Cup A	Control Cup B
Volume of water (mL)	12 mL	
Initial temperature (°C)		
Final temperature (°C)		
Temperature change (°C)		
Time bulb was on	3 minutes	0 minutes
Heat output of bulb		
Energy input to bulb	82 calories	
% waste heat produced by bulb		
% light efficiency of bulb		

QUESTIONS

1. What was the percentage of waste heat energy produced by your flashlight bulb? What do your results tell you about how efficiently these bulbs produce heat energy? Produce light?

2. Why should you use a control cup (in this experiment Cup B)? What did you place in the control cup, and what measurements did you take? Explain.

3. A typical light bulb is nearly 90% efficient at producing heat energy. Does your answer agree with the 90% figure? Why not? What problems did you have? What would you do differently?

The pie chart below shows the efficiency of a light bulb. The table gives the efficiencies of other devices.

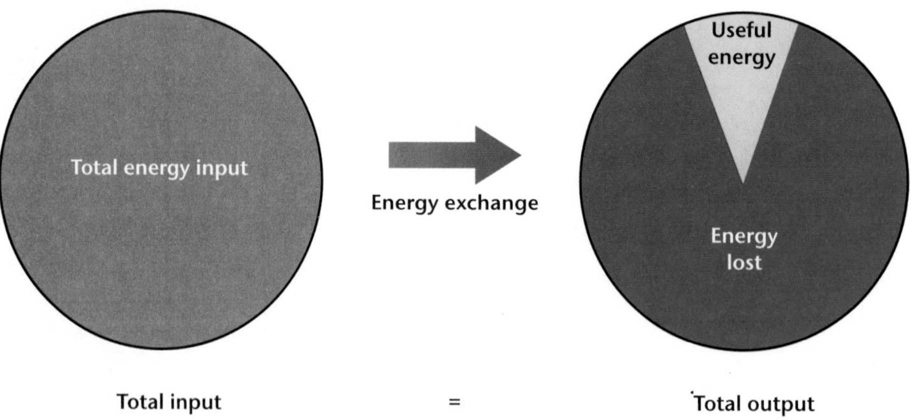

Device	% Efficiency	% Waste Heat
Fluorescent light (20-watt, 24-inch)	50	50
Incandescent light (40-watt)	11	89
Auto engine (gas)	30	70
Auto engine (diesel)	35	65
Coal power plant	38	62
Nuclear power plant	31	69
High-efficiency gas furnace	90	10
Typical gas furnace	75	25
Typical oil furnace	63	37

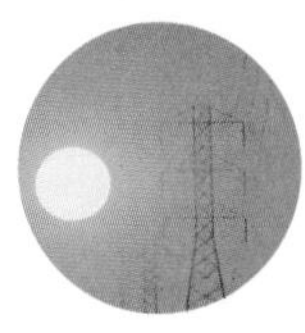

A Cool Energy Decision

Everyone uses a lot of light bulbs, but few people think much about them. When one bulb burns out, you replace it. And that's probably the last you think about light bulbs until the next time one burns out. The reading on the next page is about an important energy decision: replacing light bulbs.

CHALLENGE

Which kind of bulb should you use to replace a burned out light bulb? Incandescent or compact fluorescent? Read on and then decide.

BULBS AND TUBES

The kind of light bulb we most often use in our homes is called an incandescent (in-can-DES-ent) bulb. In this activity we discovered one of its problems: it is very inefficient. The typical light bulb produces as much as 90% heat energy and only 10% light. That means the bulb uses a lot of electricity to heat the room rather than to provide light. There is a second problem related to all the heat energy the bulb produces. Incandescent bulbs produce light by heating up a **filament** (thin wire) inside so that it glows very brightly. This filament is actually burning, and when it burns up and breaks, the light bulb goes out! Incandescent light bulbs don't last very long.

Compact fluorescent (floor-ES-ent) lamps, however, produce light in a different way. The electricity adds energy to mercury vapor inside the tube. (Scientists call this "exciting" the gas.) The mercury vapor then releases energy, which hits a special coating called phosphors (FOS-fores). This phosphor coating is spread all along the inside of the tube and converts the energy to visible light. So there is no filament, and no burning. This type of light bulb creates much less heat energy, and lasts much longer.

QUESTIONS

Note: Your teacher will help you complete a table in your science notebook that compares incandescent and compact fluorescent light bulbs. Answer each of the questions based on the information in the table. Show your work as evidence for your decision.

Comparison of Incandescent and Compact Fluorescent Lights

Type of Light Bulb	Cost to Buy	Power Rating	Average Lifetime
Incandescent		75 watts	750 hours
Compact fluorescent		18 watts	7,500 hours

If compact fluorescents are so great, why don't more people use them? Let's find out more about this energy decision.

1. Compare the cost of buying a 75-watt incandescent light bulb and an 18-watt compact fluorescent bulb. (Both bulbs provide about the same amount of light.) Based on this information, what decision would you make when replacing a burned out 75-watt light bulb? Would you buy a compact fluorescent light bulb or an incandescent one? Explain your choice.

2. Now compare the average lifetime for each type of light bulb. One compact fluorescent bulb lasts as long as how many incandescent bulbs? Using this information, what energy decision would you now make? (**Hint:** Find out the total cost of these incandescent bulbs compared to the cost of one compact fluorescent.) Explain your decision.

3. Now let's look at another factor. Let's compare the amount of electricity both bulbs use in a specific amount of time. Use 7,500 hours (the lifetime of one compact fluorescent bulb) as the specified time.

a. How many watt-hours of electricity would one compact fluorescent bulb use over its lifetime of 7,500 hours?

b. How many watt-hours of electricity would an equivalent number of incandescent light bulbs use over 7,500 hours?

To calculate watt-hours of electricity use the equation:

watts × time = watt-hours of electricity

c. Electricity is usually charged by the kilowatt-hour of use. Change your results in watt-hours to kilowatt-hours by dividing by 1,000.

4. Your teacher will tell you the cost of electricity in your area. Write it in your science notebook.

a. What will be the total cost of the electricity for the incandescent light bulbs?

b. What will be the total cost of the electricity for the compact fluorescent bulb?

(**Hint:** Use the kilowatt-hours of electricity you found for each type of bulb and multiply it by the cost of electricity.)

5. Find the total expense for using each type of bulb. Add together the cost of the bulb(s) and the cost of the electricity. Using all this information, which bulb is a better buy? Use your calculations to explain your decision.

6. What additional information would you like to know before making a final decision about which bulb to buy?

7. Which bulb do you think most people buy when they go into a store? Why? If you were a manufacturer of compact fluorescent bulbs, what would you do to get people to buy your bulbs?

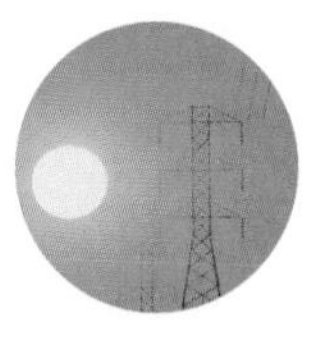

Exploring Solar Energy

The sun sends a constant amount of energy to the Earth's surface. How can we use this energy to heat water and to provide electrical power?

Construct a solar water-heating system and determine its efficiency. Then, prepare a group report describing your recommendations for redesigning the collector for maximum efficiency.

MATERIALS

For each group of four students:

1 miniature water pump
black plastic tubing
1 white plastic tubing holder—coil-shaped
1 alkaline D-cell battery
1 D-cell battery holder with leads
1 set of leads with alligator clips (one red and one black)
2 metal-backed thermometers
1 50-mL graduated cylinder
2 low-profile foam cups
1 9-ounce plastic cup
black electrical tape
masking tape
1 graph paper transparency
1 non-permanent transparency pen

SAFETY NOTE: Be careful not to break the thermometers. The metal on the thermometers may become quite hot in the sun.

PROCEDURE

1. Use the length of black tubing to construct a coil that fits within the coil impression in the plastic holder. Use black electrical tape to hold the coil together and to prevent it from moving out of the channels in the plastic holder. Be sure to leave 25 cm (10 inches) of tubing on each end free of the coil. The pump will transport the heated water to and from the cup through the ends of the tubing. Refer to the diagram below as a guide.

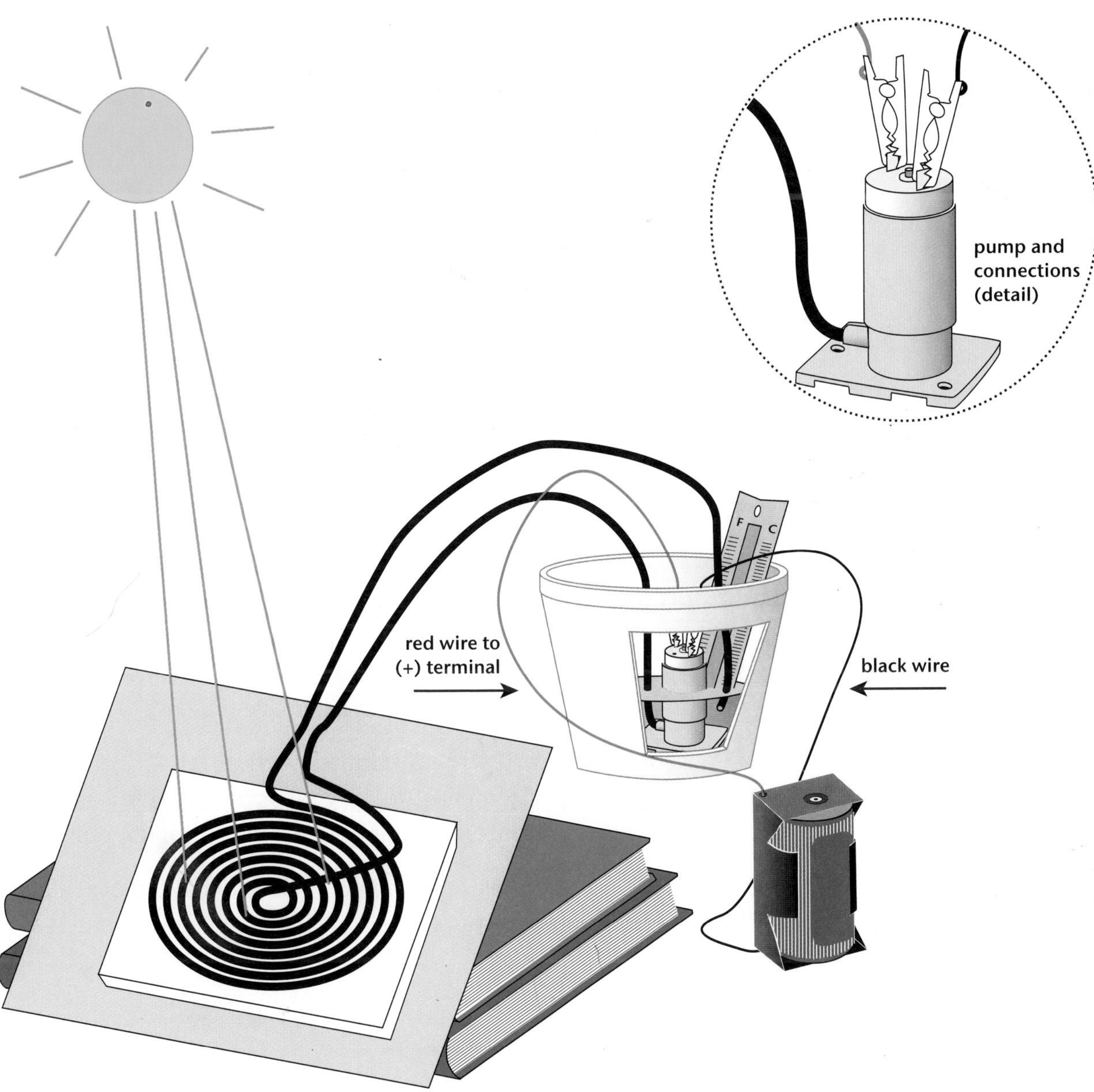

The SEPUP active solar heating system.

2. You may want to adjust the angle so that the plastic holder faces the sun. You can use a book or two to prop up one side of the holder. You may also wish to tape your collector to a piece of cardboard to give it more support.

3. Attach one end of the black tubing to the pump by pushing it into the cylindrical opening on the side of the pump's base.

4. Place the pump upright in one of the foam cups. Avoid pinching the tubing.

5. Place the other end of the black tubing in the cup.

6. Pour 100 mL of water into another foam cup. This is your control. Record the temperature of the control.

7. Place the battery in the holder so that its tip (positive end) faces the side with the red connector. Attach the red lead to the red wire, and then connect the red lead to the positive (+) terminal of the pump (look for the small plus sign). Now attach the black lead to the black wire, but don't attach it to the pump yet!

8. Measure 100 mL of water into the plastic cup. Carefully pour just enough water from the plastic cup into the foam cup containing the pump to cover the base of the pump. Save the rest to use in Step 10.

9. Take the temperature of the water in the cup with the pump. Leave the thermometer inside the cup. Do the same with your control cup. Leave both of the thermometers inside the cups.

10. To start the water pump, connect the black lead to the pump and begin your timing. As water goes from the cup into the tubing, slowly add the rest of your 100 mL of water. Do not allow the water level in the cup to rise more than 1 cm above the base of the pump.

11. Allow the pump to run for 15 minutes with direct sunlight on the tubing. Record the temperature of the water in both cups every minute. After 15 minutes, disconnect the leads from the pump.

12. Record the final temperature of the water in both the control cup and the experimental cup. Clean up as directed by your teacher.

ANALYSIS

1. Make a graph of the temperature changes over time for both the control and the experimental cups.

2. You put 100 mL of water in the cup. Calculate how many calories of energy were absorbed from the sun by the tubing and were transferred to the water.

3. Calculate the surface area of the tubing holder, which is approximately 16.5 cm by 19.0 cm.

4. Do you think you should use the whole surface area of the collector in your calculations? Why or why not?

5. If the sun supplies 1.5 calories of energy per square centimeter per minute to the Earth, how many calories were supplied to your collector for 15 minutes?

6. Now calculate the efficiency of the solar heating system. The efficiency would be the number of calories actually gained by the water in the system divided by the total number of calories supplied to the system by the sun.

7. As a group, write a short report describing the outcomes of this experiment. Include the graph and calculations you completed above. Also include suggestions on how you would redesign your collector to increase its efficiency.

Using Solar Energy

The sun is the ultimate source of the energy we use. It is also a renewable energy source. In this reading you will learn how this abundant source of energy can be used to heat swimming pools.

Use the information from this reading to help you understand how to construct a better hot-water collection system using the sun's "free" energy.

WARMING WATER WITH THE SUN

The energy costs of heating an outdoor swimming pool with an electric or gas water heater are very high. What's worse, the energy used to heat the water in the pool quickly transfers to the air around it.

Fortunately, outdoor swimming pools are warmed passively by the sun when the sun's radiant energy falls on them. (*Passive* is the opposite of *active*.) This warming by the sun is called passive because you do not have to do anything special to the pool to warm it. Some of the sun's energy is transferred to the water and heats it. Some is reflected off the surface into the air. Of course, in many parts of the country passive solar heating does not warm a pool enough to get most people to dive in, at least not in the morning!

Pool covers reduce heat loss and increase safety.

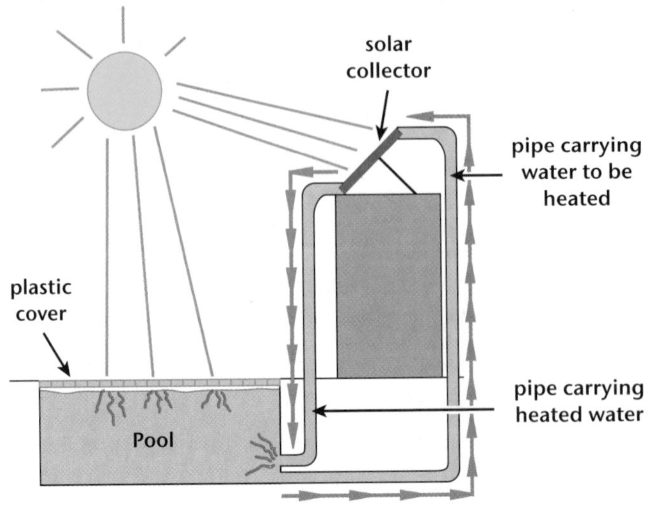

An active solar water heating system

There are several kinds of solar heating systems for pools. They are used to increase the amount of energy transferred to the water to heat it. A total system not only heats the water, it helps prevent the loss of energy to the outside environment. Preventing energy, or heat, loss is usually done by one or a combination of two methods. One method, called a passive system, places a clear plastic cover with air cells in it over the pool. This system allows some light energy to pass through the cover to heat up the water, but the cover also acts as a heat insulator to slow down the rate of heat loss to the air. Using such a cover at night, or when it gets cool, is especially effective at reducing energy costs.

The second system, called an active system, involves placing black plastic panels with water-flow channels in them on the roof of a nearby building in a specially constructed holder. They are much like the coils you used to build a water heating system in Activity 9. The panels are aimed towards the sun for as much of the day as possible. Water pumped from the pool up to the panels flows through the channels. The black plastic is an excellent absorber of radiant energy, so the water flowing through the panels is warmed and returned to the pool.

This is a closed system because when the sun is shining, all of the water in the pool automatically flows through the panels as part of the pool's water circulation system. Heat sensors are used to automatically turn the system on or off whenever the water standing in the channels reaches a certain minimum or maximum temperature. This means that the water is pumped through the channels only when they are warm.

Although an active system uses electrical energy to pump the water up to the panels, this same pump is already used to circulate and filter the pool water. A well-designed active system will heat the pool more effectively than a passive system made up of just a cover. Combining systems both provides heat from the active system when the sun is shining and limits heat loss at night when the pool is covered.

QUESTION

What are the main differences between passive and active solar water heating systems?

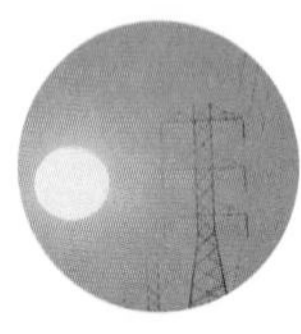

The Photovoltaic Effect

Investigate how a solar cell can directly transform solar energy into electricity.

Explore the use of solar cells to produce electricity.

Photovoltaic array near Albuquerque, NM

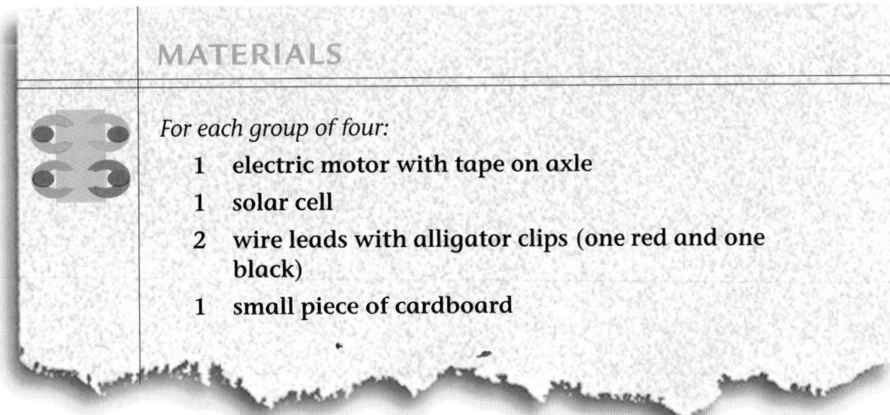

MATERIALS

For each group of four:

1 **electric motor with tape on axle**
1 **solar cell**
2 **wire leads with alligator clips (one red and one black)**
1 **small piece of cardboard**

PROCEDURE

1. Work together in your group to set up and test the solar electric cell. Use the diagrams on this and the following page as a guide. Describe the various conditions—angle to the sun, amount of sunlight, electrical connections—that produce the fastest spin of the motor. How can you measure the rate of spin?

2. Now join with another group to connect two solar cells together to drive one motor. Your goal is to furnish more energy (voltage) to the motor. As a group, decide how you might go about doing this and try it out. Discuss your results with other groups.

3. To simulate the effect of a cloudy day, use the piece of cardboard to block off first one-quarter, then one-half, and finally three-quarters of the surface of the two solar cells. Keep a record of what effect this has on the speed of the motor.

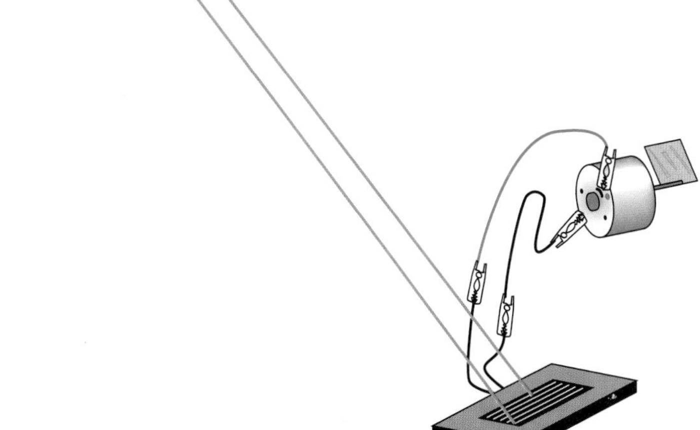

ANALYSIS

Summarize in your science notebook what you have learned about solar cells. Include a short statement recommending or rejecting solar cells as your only source of energy. Be sure to give evidence for your reasoning.

GOING FURTHER

If your teacher permits, try to run the pump for the solar water-heater system using electricity from the solar cells rather than the D-cell battery. Will one solar cell power the pump, or does it take several? With just the battery, you were not able to control the speed of the pump. By positioning or covering the solar cells you can control the speed. Can you use the solar cell to determine at what rate of pumping the water heats more quickly? Should the water circulate quickly or slowly? Working as a group, develop a written report to describe what you did and what you found out.

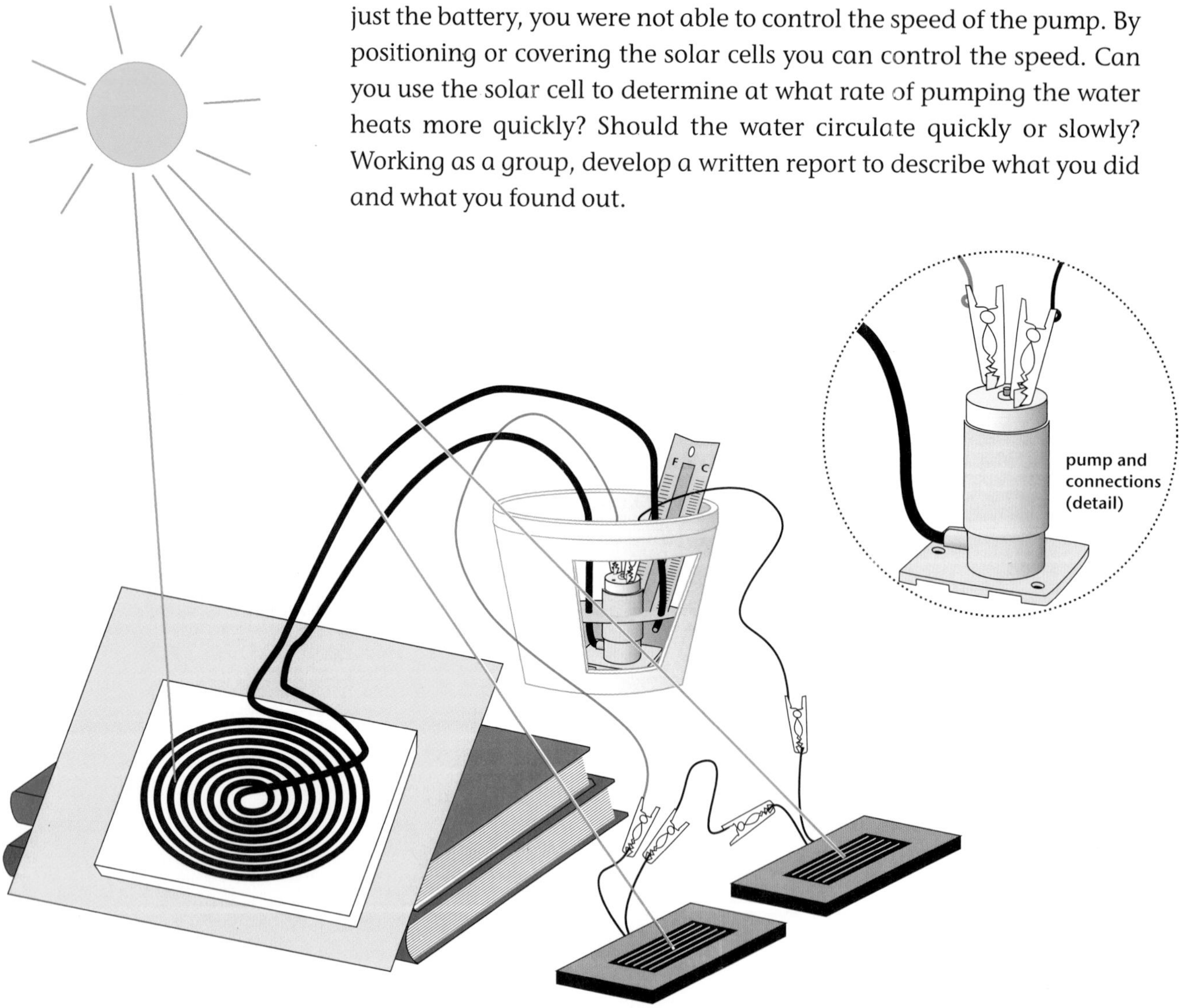

pump and connections (detail)

Using photocells with the SEPUP active solar heating system

10 Controlling Radiant Energy Transfer

A "Reflective" Question

In the last activity, you investigated how to capture and store energy from the sun. Sometimes, the sun provides heat where or when it is not wanted. In this activity, you will look at methods for preventing unwanted heating by the sun.

Learn how the transfer of energy from the sun can be controlled by special materials.

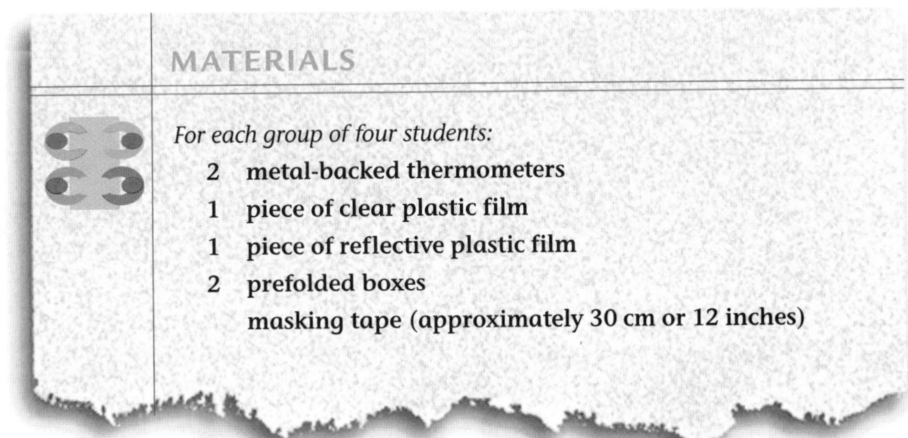

MATERIALS

For each group of four students:

2 metal-backed thermometers
1 piece of clear plastic film
1 piece of reflective plastic film
2 prefolded boxes
masking tape (approximately 30 cm or 12 inches)

PROCEDURE

Examine your materials and your two pieces of plastic film. Plan an investigation that will allow you to compare the ability of these two plastic films to prevent the transfer of the sun's heat through a window into a room. Use the box and thermometer as shown below to simulate a room with a window. Carry out your investigation.

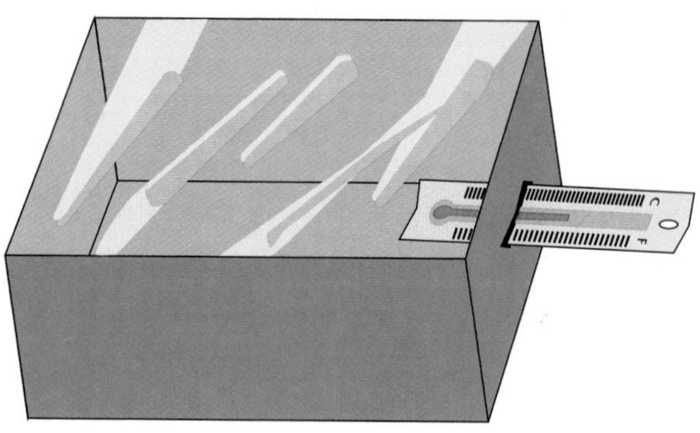

ANALYSIS

1. Write a report describing the results of your investigation.

2. Think about the investigations you did in the last two activities. There were some similarities and some differences between the investigations. Describe how you used materials to accomplish the different energy transfer goals in Activities 9 and 10.

   Be sure to identify:

   a. the energy chains involved;

   b. the effect of the materials chosen on the energy transfer; and

   c. the results of the decisions made in each case.

Driving Trade-offs

Automobiles—and their emissions—have changed the way we live. What impact will they have on future generations?

CHALLENGE

Read about Mexico City today and in a possible, imagined future. Assuming that cars are a major source of pollution, could this future become a reality some day in large cities across America?

Rush hour traffic in Washington state

LIVING WITH AIR POLLUTION

March 9, 2033

John's hovercraft shot smoothly over the black volcanic peaks, and then set a straight course for the deserted city ahead. It had been nearly 15 years since he had last been here, in 2018, just before the massive evacuation. Now the city, with its tall towers and once-green parks, busy roads, and cluttered markets, sat empty against the horizon of brown air. He had a few moments before he had to put on his gas mask. He remembered Ricardo's face when Ricardo and his family had to flee the city. Where was Ricardo now? John looked down at the old journal sitting in his lap and opened it to a page he had written in 2016. He began to read...

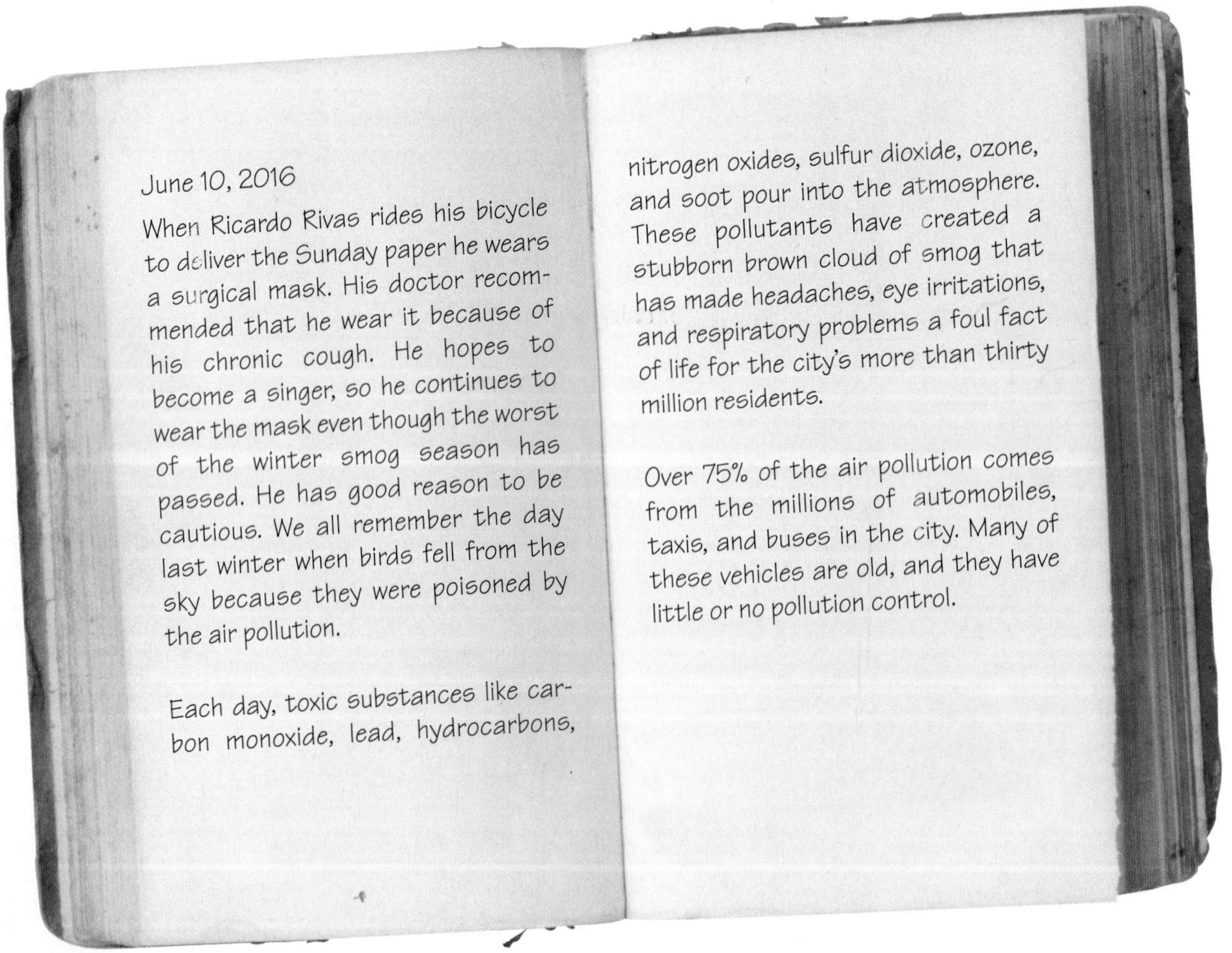

June 10, 2016

When Ricardo Rivas rides his bicycle to deliver the Sunday paper he wears a surgical mask. His doctor recommended that he wear it because of his chronic cough. He hopes to become a singer, so he continues to wear the mask even though the worst of the winter smog season has passed. He has good reason to be cautious. We all remember the day last winter when birds fell from the sky because they were poisoned by the air pollution.

Each day, toxic substances like carbon monoxide, lead, hydrocarbons, nitrogen oxides, sulfur dioxide, ozone, and soot pour into the atmosphere. These pollutants have created a stubborn brown cloud of smog that has made headaches, eye irritations, and respiratory problems a foul fact of life for the city's more than thirty million residents.

Over 75% of the air pollution comes from the millions of automobiles, taxis, and buses in the city. Many of these vehicles are old, and they have little or no pollution control.

John put the journal down and thought, "Why did this have to happen? Was there any way that people could have changed their cars or their driving habits to save the city?" He thought again about Ricardo as he climbed out of the hovercraft. The deserted city lay silent, dead.

Does the story you just read sound impossible? Programs aimed at reducing vehicle emissions have been developed in the hope of avoiding a scenario like the one just described. However, it will take years before residents of some of the world's largest cities can be assured of clean air.

Mexico City currently faces extremely high pollution levels and is trying hard to reduce them. Its taxis and buses are being replaced or updated with new engines. New cars must use unleaded gasoline, and catalytic converters became standard equipment on new cars in 1991. With a population of over 26 million people, Mexico City is congested with vehicles. The situation is made worse by the city's high altitude location at 2,240 meters (7,347 feet). Fuels do not burn as efficiently there as they do at sea level because the atmosphere contains less oxygen. This means cars produce more pollutants. In addition, the city is surrounded by mountains that trap the polluted air near the ground during temperature inversions. A temperature inversion occurs when a layer of warm, sometimes polluted, air is trapped near the ground by a thick layer of colder air. This usually happens in the winter in valleys

A smoggy summer day in Los Angeles

or other areas where air is trapped. Usually, these inversions last only a few hours, but people in Mexico City fear longer inversions. Long temperature inversions helped cause the death of 20 people in the town of Donorra, Pennsylvania in 1948, and 4,000 people in London, England in 1952.

The United States began to require its cars to be less polluting long before Mexico City started, and the air in some major US cities has become cleaner as a result of pollution control. Nevertheless, according to 2001 United States Environmental Protection Agency (EPA) estimates, more than 64 million people live in areas with "extreme" ozone risks, and more than 17 million live in areas with "serious" risks from carbon monoxide. Both ozone and carbon monoxide are common air pollutants that result from automobile use. Reducing pollution from automobiles will have to be included in any plans to further clean up our cities' air.

QUESTIONS

1. Why does Ricardo wear a surgical mask?
2. What are some of the major causes of air pollution in Mexico City?
3. Assuming that cars are a major source of pollution, suggest a possible plan for preventing the conditions described in the story about the future. Then describe what you would need to do to find out if your plan is practical.

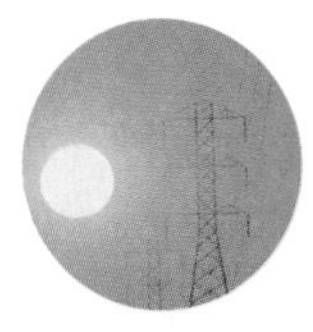

Automobile Fuel Efficiencies

Imagine you are about to buy your first car. What model do you think you will choose? A lot of factors will probably influence your decision. One factor you may consider is the cost of buying the car. You may also consider the fuel efficiency in miles per gallon (mpg). The greater the fuel efficiency, the less you will have to pay for gas.

CHALLENGE

Prepare graphs of the data shown in the table on the next page and use the graphs to understand factors that affect fuel efficiency.

Mileage Estimates for Selected Cars				
Car Name and Model	**Weight (lbs)**	**Horsepower**	**Number of Cylinders**	**Fuel Efficiency (miles per gallon**)**
Acura RSX*	2,780	160	4	27/33
Buick Park Avenue	3,640	205	6	20/29
Buick Regal	3,445	160	6	18/27
Chevrolet Lumina	3,395	160	6	21/32
Chrysler Concorde	3,605	214	6	20/28
Ford Focus	2,715	110	4	28/32
Ford Taurus	3,516	145	6	20/28
Dodge Neon*	2,559	132	4	24/31
Geo Prizm*	2,510	105	4	29/33
Honda Accord	3,255	130	4	23/30
Honda Civic	2,440	106	4	30/34
Lincoln Town Car	4,055	210	8	17/24
Nissan Altima	3,050	150	4	23/29
Chevrolet Blazer, 2WD	4,225	190	6	16/22
Ford Expedition 4WD	5,290	240	8	14/17
Saturn SL	2,405	100	4	27/37
Suzuki Esteem GL*	2,290	95	4	27/34
Toyota Corolla	2,540	100	4	29/33

*manual transmissions

**based on 2001 EPA data, city/hwy

PROCEDURE

1. Work with your group of four to graph the data. One pair of students should make a graph that compares the fuel efficiency of each car to its weight. The other pair should compare the fuel efficiency to the horsepower. For both graphs, plot fuel efficiency on the y-axis.

2. Share your graph with the other pair in your group as you answer the questions.

ANALYSIS

1. Describe the trends (patterns) shown by your graphs. Which variable—weight, or horsepower—seems to have the greatest effect on mileage? Does the number of cylinders make a difference?

2. Which one of these cars would you choose to buy? Why? What more would you like to know before deciding?

The Energy Efficiency of Modern Cars

Automobiles have changed rapidly in style and energy use since the earliest models. But, as you can see from the graphs on the following pages, their gasoline efficiency has made little progress since 1989. Why do you think this is? How can we improve cars and make them more efficient?

CHALLENGE

After discussing the following information with your group and family, develop a list of the most important things to consider if you were to design a new car, the "car of the future."

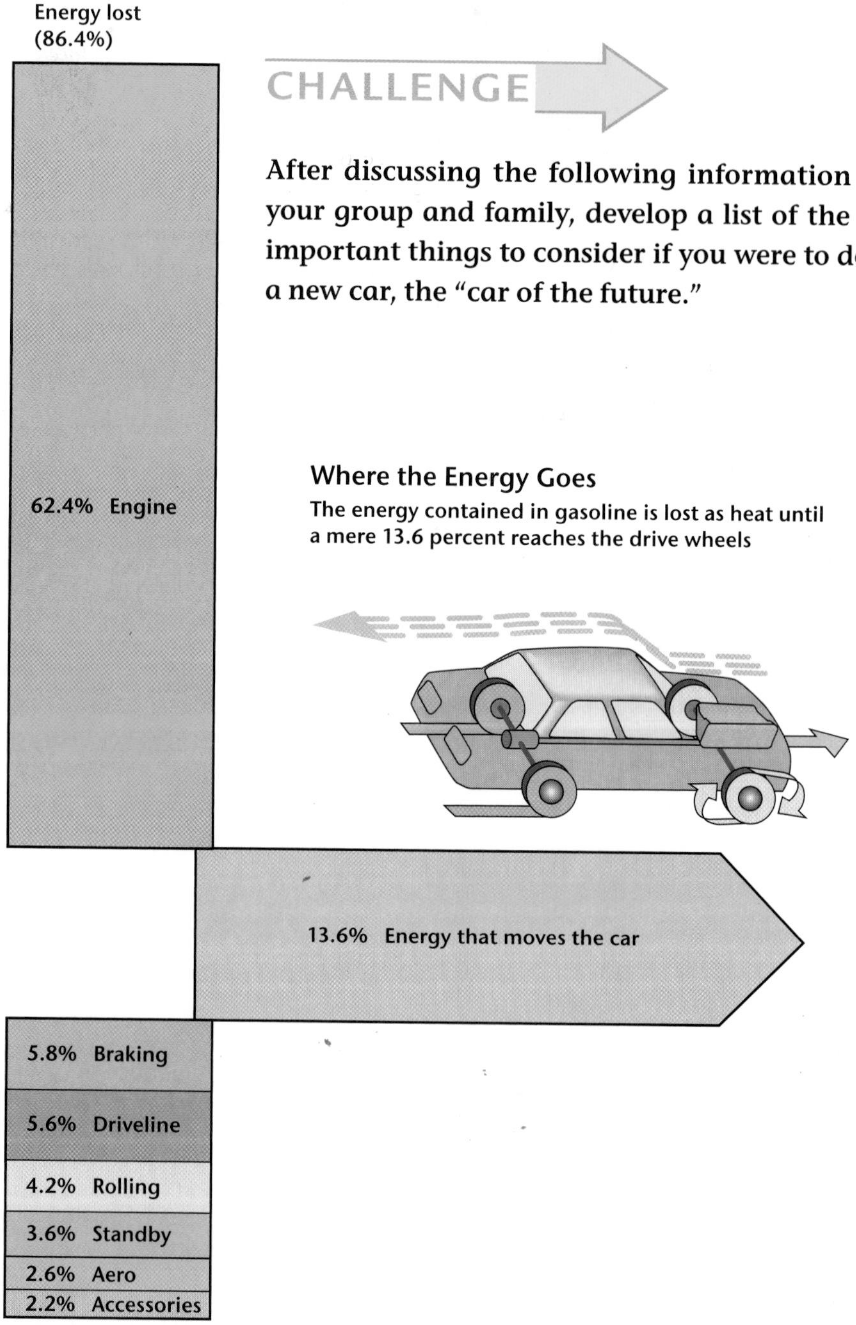

Where the Energy Goes

The energy contained in gasoline is lost as heat until a mere 13.6 percent reaches the drive wheels

Energy lost (86.4%)

62.4% Engine

13.6% Energy that moves the car

5.8% Braking

5.6% Driveline

4.2% Rolling

3.6% Standby

2.6% Aero

2.2% Accessories

Energy Sources—Advantages					
	"Clean" Gasoline	Ethanol	Compressed Natural Gas	Electricity	Hydrogen
Description	Today's gasoline combined with oxygen-giving compounds	Alcohol fuel made by fermenting corn and sugar cane	Used for cooking and heating in homes—methane	Derived from recharging a variety of new, high-powered batteries	Solar energy would be used to split water molecules into hydrogen gas. This would be used in a fuel cell in the car to create electricity.
Advantages	Reduces air pollution from all cars that use it Can be distributed now in present gas stations Doesn't require engine changes	Has high octane (105) compared to regular gas (87) Comes from a renewable resource Has lower pollutant levels (carbon monoxide) than gasoline	Is abundant in the U.S Lowers pollutant levels emitted by 60% on the average Has distribution system already in place	Has quiet engine Emits almost no pollutants at the point of use	Comes from a renewable energy source Emits practically no pollutants Produces no carbon dioxide

Energy Sources—Disadvantages					
	"Clean" Gasoline	**Ethanol**	**Compressed Natural Gas**	**Electricity**	**Hydrogen**
Disadvantages	Benefits to the environment are not known Uses a non-renewable resource Does not reduce carbon dioxide emissions	Produces less energy per gallon than gasoline Requires frequent fill-ups Would use 40% of U.S. grain harvest to produce only 10% of fuel needed	Requires heavy fuel tanks Requires refueling every 100 miles Takes two to three times longer than gasoline to refuel Comes from a non-renewable source	Technology not yet widely available Requires over 1,000 lbs. of batteries that need to be recharged for 6–8 hours daily; range is limited to 100 miles The pollution would be transferred to the source of electrical generation. Speed limited to 30–65 mph	Technology not yet widely available Cannot match the performance of today's cars No distribution system is in place to ship the gas. There are safety issues of hydrogen gas explosions.
Costs	12–14 cents more per gallon than present unleaded gasoline	Twice the cost of gasoline	At the moment inexpensive—70 cents per gallon	Cost of electricity to recharge batteries is low, but batteries must be replaced every 30,000 miles at a cost of $2,000 each	Very expensive and not currently produced on a large scale

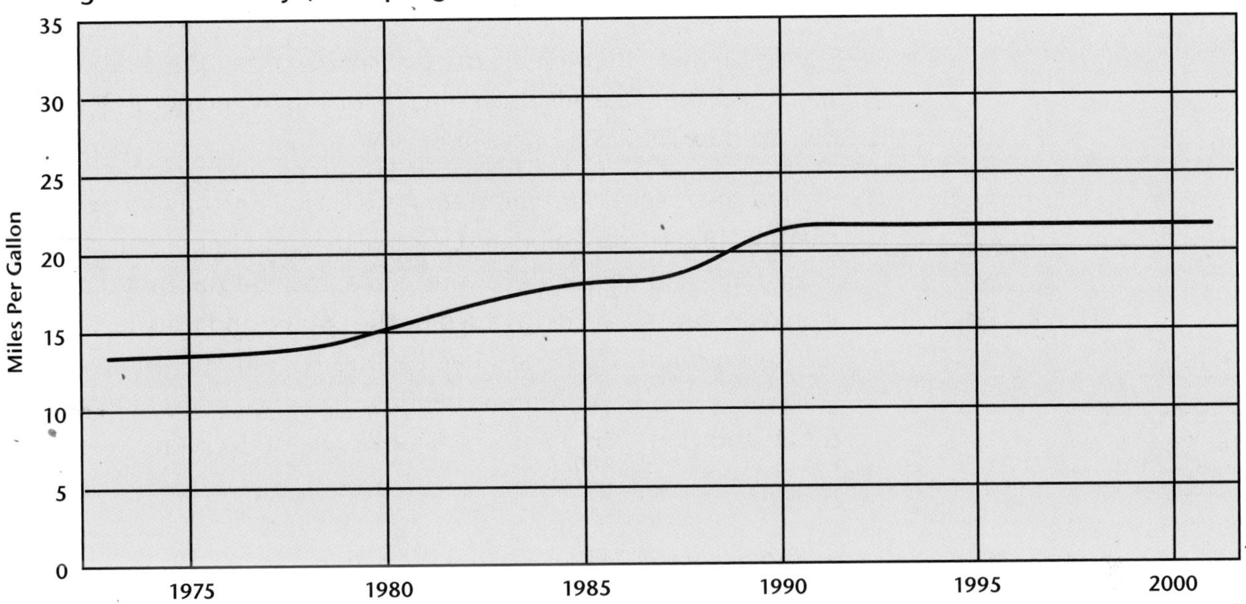

Changing the Efficiency (Fuel Economy) of the Modern Car	
Change	**Approximate Effect on Efficiency**
Adding an air conditioner	**reduces** efficiency by 9%
Using an automatic rather than manual transmission	**reduces** efficiency by 6%
Using steel-belted radial tires	**increases** efficiency by 10%
Reducing driving speed from 65 to 55 mph	**increases** efficiency by 5%
Reducing the vehicle weight by 500 lb.	**increases** efficiency by 15%

QUESTIONS

1. Collect an advertisement from a newspaper or magazine about any new car. What reasons do you have for buying or not buying this car? Be prepared to share your reasons in class.

2. Design a new car that is more energy efficient and environmentally friendly than current models. Describe the most important changes you would make to achieve your goals. Include the materials you would use and why, energy used by the engine and the fuel it might use to power it, and the effect of the car on the environment (including effects on human health). Be sure to describe the evidence and trade-offs you used in making your decisions.

   Include:

   - the materials
   - energy used

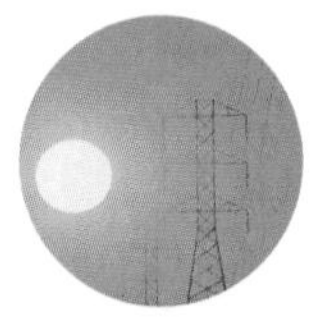

The Hypercar

In 1993, leaders from government and the automobile industry held an historic conference. Together they proposed the goal of manufacturing super efficient, environmentally friendly cars by 2005.

The proposed cars will have fuel efficiencies of up to 80 miles per gallon of gasoline.

CHALLENGE

Read about the hypercar and compare it to the car of the future that your group has been designing.

The Toyota Echo (left) and Prius (right) share the same chassis and body. Yet the Prius delivers much better gas mileage. Why do you think this is so?

HYPERCARS

Materials scientists, automobile engineers, energy specialists, and other experts are continuing to develop a car that is completely different from the cars of yesterday. These cars represent a complete breakthrough in material and energy use. They are called "hypercars."

A typical car loses lots of energy through heat loss, friction, and air resistance. For every dollar of energy that goes into the gas tank, only 14 cents comes out as usable energy to move the car. The rest, 86%, is wasted! But if new thinking about materials usage, energy storage devices, and **aerodynamic** (air-oh-die-NAM-ic) design, together with high-tech manufacturing processes, are applied to this problem, a truly remarkable vehicle can be produced. (An aerodynamic design allows air to flow over the body of the car with the least resistance, reducing the amount of energy needed to push the car through the air.)

Most of the present cars' weight comes from the steel body and the engine. With the introduction of synthetic materials—carbon fibers that are stronger and much lighter than steel—the car's body weight can be cut from an average of 3,200 pounds down to about 1,400 pounds. This change alone has a big impact. It means that the engine can be much smaller and need less horsepower. In fact, the engine might need to be only slightly larger than that of a large lawnmower! That would result in greatly increased fuel efficiency.

Hypercars have two energy sources. Gasoline or an alternative fuel such as ethanol, natural gas, or hydrogen is the primary fuel, but a generator linked to the engine creates electricity that powers electric motors attached to each wheel. When the car slows down or brakes, the generator also recovers some of this energy and stores it in special large batteries for future use. Hypercar hybrids like the Honda *Insight* and Toyota *Prius* are far more fuel-efficient than conventional cars, offering EPA mileage estimates in the 50–60 mpg range.

New developments in technology are occurring all the time. New kinds of engines, batteries, and materials technologies are still in development, but much of this new technology is still quite expensive. By molding the frame and body into a single carbon fiber unit, hypercar designs minimize the amount of energy lost to air resistance. Carbon fiber, however, typically costs 30–40 times as much per pound as steel, which means hypercar frames and bodies are much more expensive to produce than those of conventional cars. Hydrogen fuel cells, which may power future hypercars, use an inexpensive and abundant fuel source—water—have higher efficiencies, and produce much less air pollution. However, the fuel cells themselves are costly, and a car powered by them would be beyond the reach of most Americans' pocketbooks. It would also be smaller, and all of us would need to change the way we view cars. Future hypercars will make use of these and other important advances as the technology develops and costs come down. A car that can cross the country on one tank of gas may not be that far in the future.

Improving hypercar design and performance at a reasonable cost will be a great challenge. Meeting this challenge is likely to help solve other problems. The new technological advances and materials developed in producing hypercars are likely to have many other uses, not only for cars but in our homes and other aspects of our lives.

QUESTIONS

1. Compare your car design with the hypercar. What new ideas would you like to add to your design?

2. What if the hypercar were made so inexpensively that every person in the world could own one? Write three or four paragraphs explaining your reaction to a world with as many cars as people. Include your ideas about alternatives to this future.

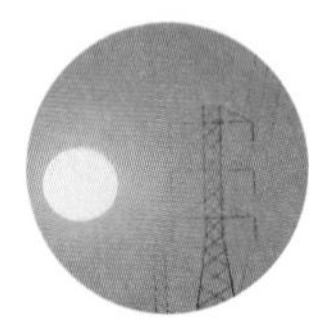

Comparison Shopping

	Toyota Prius	Toyota Echo
Manufacturers Suggested Retail Price	**$19,995***	**$15,525***
Engine type	Gasoline and electric	Gasoline
Horsepower	70@4,500 rpm	108@6,000 rpm
0–60 mph acceleration (seconds)	12.8	9.5
NHTSA frontal crash safety rating (% chance of life-threatening injury)	20%	15%
Head room (in)	38.8	39.9
Weight (pounds)	2,765	2,100
Seating capacity	5	5
Trunk volume (cubic feet)	12	14
Fuel tank volume (gal)	11.9	11.9
City/Highway Fuel Efficiency (miles per gallon)	52/45	32/38
Cost of driving 15,000 miles**	$469	$608
Miles between fill-ups	571	440

*Comparably equipped

**Calculated using combined city/highway mileage and $1.50 per gallon as the price for gasoline

QUESTIONS

1. What are the advantages of the *Prius* over the *Echo*?
2. What are the advantages of the *Echo* over the *Prius*?
3. How many miles would you have to drive to make up the difference in price between the *Prius* and the *Echo*?

12 Designing an Energy-Efficient Home

In Activities 5 and 6, you investigated electrical energy use in your home. You made a plan for reducing the amount of electrical energy you use. There are other ways to approach energy efficiency at home. Today, many architects and home builders are turning to energy-efficient designs and solar power to help reduce the use of electricity and gas for heating and cooling homes. The approaches they use include selecting energy-efficient building materials and techniques, carefully locating and orienting the home relative to the sun and the site, and using passive solar heating and cooling.

CHALLENGE

Design an energy-efficient house for your local climate that meets the needs of your family. The house should be warmed by the sun in winter but not overheated in summer. It may also take advantage of other energy-efficient methods you have learned about in this book.

PASSIVE SOLAR ENERGY AND THE HOME

Energy-conscious design is a growing field in the building profession. Taking advantage of the sun, climate, and local materials to provide a comfortable dwelling is not a new idea. Consider the materials used to construct the thick-walled Spanish missions in the Americas, the cliff dwellings of Native Americans in the Southwest, or the underground homes built in the desert of Australia. Each of these types of home takes advantage of materials and design principles to maintain a comfortable temperature without relying on mechanical heating and cooling equipment.

Passive solar heating can be accomplished simply by leaving the south side of a house unshaded during the winter, by using wall materials that store heat, or by building sunrooms. Passive solar cooling can be accomplished by ensuring that there is good ventilation, and by providing shade during the hot summer months. Energy-efficient homes may take advantage of the waste heat from incandescent lighting systems and other appliances or minimize the heating effects associated with their use, depending upon the climate and season. Solar heating can also be used to provide hot water, as you saw in Activity 9. In Activity 10 you used some of the new kinds of window coverings that control the transfer of solar energy. Think about how you can use these ideas in your planning.

Solar panels on a roof.

PROCEDURE

1. Prepare a drawing of an energy-efficient house for your local climate that meets the needs of your family. Label your drawing. You may wish to write a brief description of your design and why you think it will reduce the use of electricity and/or gas (approximately 200–300 words). Be sure your design uses some of the approaches you have learned about in this part of the course. You should also include each of the following:

   - Climate
     Include the sun's position and seasonal movements, wind direction, and variations in temperature and humidity.

   - Orientation of the house
     Keep in mind the direction that each of the rooms faces.

   - Windows and shading
     Describe the direction of major windows and the use of shades, eaves and/or landscaping.

   - Building materials
     Choose materials that absorb warmth from the sun and can release the warmth gradually during the winter, and be sure to provide insulation against both summer heat and winter cold.

2. If you have time, make a model of your design and test it in the sun or with a light bulb as a simulated sun.

ANALYSIS

Examine plans and strategies used by different students in your class. What changes would you make based on your comparison?

GOING FURTHER

Many local energy utilities provide information describing steps homeowners can take to reduce energy costs. Obtain this information and use it to improve your design.

Environment

CambridgeSide
GALLERIA

Environment

In this part of *Issues, Evidence and You,* you will have an opportunity to apply what you have learned in earlier parts of the course to a new issue—whether or not to build a factory in an island community. People are probably making decisions like this in your community right now. They may be deciding whether to build industrial plants, shopping malls, a housing development, or a new transit system.

When a community is considering a major development project, the agencies involved usually prepare an Environmental Impact Report, often referred to as an EIR. The decision about whether an Environmental Impact Report is required is made at the local government level when there is evidence that a project will have a significant impact on the environment. The purpose of an EIR is to tell the public how the project will affect the environment and what steps will be taken to reduce the impact. The benefits of the project are also reported. The decision about whether to approve a project is not based on a threshold of impact, that is, on whether it will cause a given amount of harm to the environment, but rather on an overall balance of the impacts and benefits of the project. In other words, if the benefits outweigh the negative impacts, the project may still be approved.

You will play the role of a member of a team that is trying to plan a factory that will benefit an island community while causing as little negative impact on the environment as possible. You will first learn about the factory's product, including its method of production and the wastes it generates. Each team will prepare a factory proposal, including an Environmental Impact Report, for consideration by the island's residents, who will be played by your classmates. Your report will include your plans for obtaining the water, materials, and energy the factory needs, disposing of the factory wastes, and maintaining water quality and the overall environment of the island. When all the teams have completed their reports, the class will decide whether to approve one of the factory proposals.

Comparing Four Industries

How can members of a community compare the possible impacts of four different industries? You will begin to investigate this process by rating four industries based on what you already know about them. Then you will gather more evidence about the products and by-products of each industry before reconsidering your ratings.

CHALLENGE

Rate four industries according to the categories in the chart on page 6. Identify the advantages and disadvantages of locating each industry in your community.

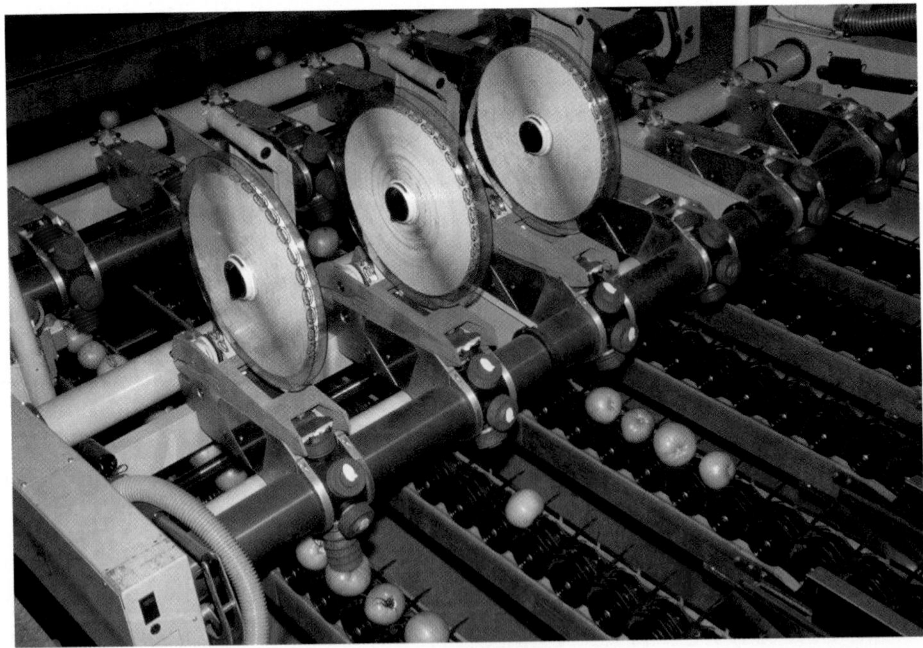

Tomato processing plant

INDUSTRY DESCRIPTIONS

Chemical manufacturing

Many chemical companies specialize in specific types of chemicals. Consider a chemical industry that makes plastics for a wide range of products. The raw materials used in making plastics come from processed crude oil. Energy and some chemical materials are needed to transform the raw materials into plastic, and waste by-products are produced. The plastic is usually shipped to other factories to be manufactured into final products.

Computer production

Manufacturing computer circuit boards requires energy, special plastics, cleaning solvents, and other materials, such as acids. Various chemical processes are used to produce the circuit boards.

Food processing

Raw tomatoes are brought to processing plants where they are washed, crushed, cooked, and made into a variety of tomato products. They are packaged for shipment to markets. The process requires the use of some chemicals in addition to the natural ones in the tomatoes.

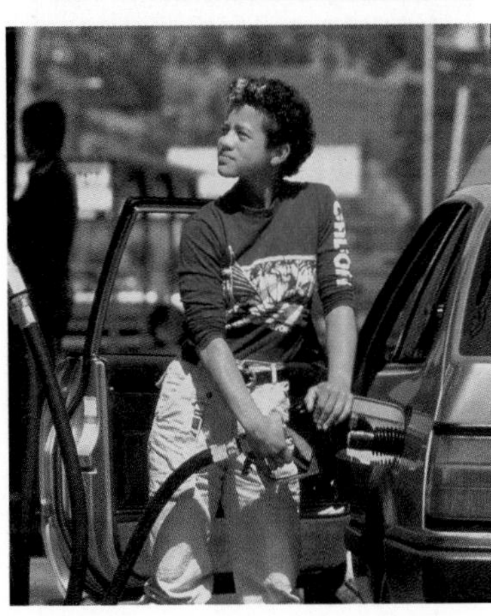

Oil refining

Crude oil comes into a refinery by pipeline. Energy and water are required to process it to produce gasoline, other fuels, and the raw materials used for making plastics.

PROCEDURE

Part One: Comparing Four Industries

1. Read about the four industries that are briefly described on the previous page. Imagine that they want to locate in or near your community.
2. In your science notebook make a chart similar to the chart shown below.
3. Rate each kind of industry on your chart, using the 1–5 scale for each pair of word opposites. For example, in section C, if you think an industry is valuable, mark a 5 in the appropriate space. If you think an industry is worthless, mark a 1. Use a 3 if you think the industry is neither valuable nor worthless. Use a 2 if you think it is somewhat worthless or a 4 if you think it is somewhat valuable.
4. Base your ratings on the information about each industry given on the previous page and on anything else you know about the industry.

Ratings

	Category 1 2 3 4 5	Chemical Manufacturing	Computer Production	Food Processing	Oil Refining
A	bad good				
B	ugly beautiful				
C	worthless valuable				
D	dirty clean				
E	harmful helpful				
F	hazardous safe				
G	unnecessary necessary				
	My Totals				
	Our group's average				
	Class average				

PROCEDURE

Part Two: Evaluating Industry Siting Issues in Your Community

1. Imagine that your community is trying to decide whether to encourage new industry to come to town. You have been asked to join a citizens' committee that is evaluating the advantages and disadvantages of bringing new industry to the community. In preparation for the committee meeting, make a chart of all the advantages and disadvantages you can think of in locating an industry in your community.

2. Compare and discuss the charts of all members of your evaluation team. Reach a consensus on the three most important issues to resolve before a decision can be made about whether an industry can come to your community.

3. As a class, come to agreement on the three most important issues. Record them in your science notebook.

Evaluating Industry Reports

Your team will evaluate detailed reports submitted by four industries that want to locate in your community. Each team member will produce a summary about one industry and share the results with the other members of the team.

CHALLENGE

Read the overview report submitted by each industry. Separate the evidence from the company's opinions about its own operations and safety.

Computer chip manufacturing

PROCEDURE

1. Read about your assigned industry. As you read, try to separate evidence from opinions.
2. In your science notebook, prepare a summary about the industry that includes answers to the following questions.
   a. What are the raw materials (reactants) used in the industry?
   b. What are the various products of the industry?
   c. What are the by-products of the industry? Are they reused or disposed of as wastes?
   d. What are three *advantages* of having this industry in the community?
   e. What are three *disadvantages* of having this industry in the community?
3. Share your results with the other members of your group.

INDUSTRY OVERVIEW: KELLY CHEMICALS

Working for You

Kelly Chemicals owns and operates 15 plants around the country. These plants produce many chemical products that most people use daily in one form or another but may not know by name. Kelly Chemicals makes substances that are used in textiles, lubricants, tires, plastic packaging, detergents, adhesives, and cosmetics. Kelly Chemicals would like to build a plant in your community, providing employment and needed products for the local economy.

Making Chemical Building Blocks

The proposed plant's main products—olefins—are compounds of carbon and hydrogen made from natural gas. They are sometimes called "building blocks" because they must be processed further before they can be turned into finished products. Some of the major types that will be produced at this plant are ethylene, propylene, polyethylene, and polypropylene. These molecules are the raw materials that are used to produce many common consumer products. Other chemicals made at the plant will be sold to other chemical plants for additional processing into such items as dry-cell batteries, plastics, detergents, adhesives, and synthetic oils.

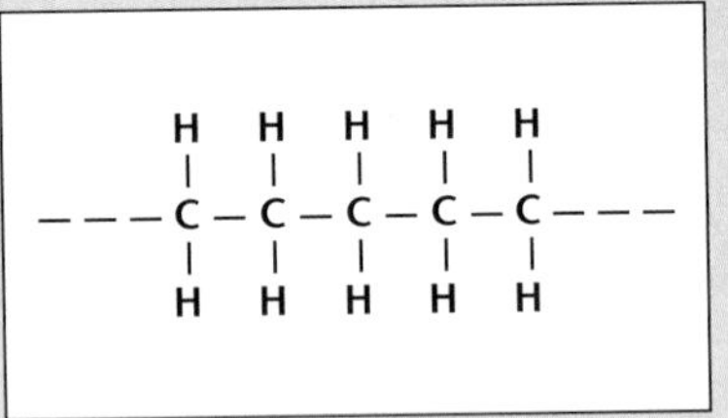

Part of a hydrocarbon chain

Community Concerns

We know that there are many concerns regarding the operational safety of the plant. Other plants of similar design and operation have earned numerous safety awards. Such plants have chalked up more than 2 million continuous people-hours without an accident serious enough to require time off from work.

Fire and Explosions

The olefins, and the hydrocarbons used to make them, are flammable. Sophisticated electronic and mechanical means will be used to monitor operations for signs of problems. The ethylene unit alone will have 1,300 sensors that continuously relay information to a control room where engineers can, if necessary, shut down operations in the plant. We will regularly test our detailed emergency procedures, which will be developed in cooperation with local community leaders. Employees will be required to keep their skills sharp with regular training sessions and updates.

Waste Issues

Kelly Chemicals has a policy of waste minimization. Plant managers work with engineers to develop processes that minimize the waste generated by plant operations. We register all solid waste by-products with appropriate authorities and dispose of them in compliance with all federal, state, and local regulations.

We treat wastewater from plants, like the one proposed for your community, to a level of quality far higher than required by state laws. Processed wastewater from one of our other plants supplies a well-stocked artificial lake rich in native waterfowl, fish, and other animals. A similar project will mitigate (compensate) for the impact of the proposed factory. We will construct an artificial lake near the proposed factory site that will provide sanctuary for local birds affected by the industrial area.

A Community Benefit

The many uses of olefins ensure their continued importance to the economy. The plant will provide jobs for more than 700 people. The plant will contribute millions of dollars in local, state, and federal taxes, and many more times that in local salaries.

Kelly Chemicals believes it can operate a plant that is safe for workers and community residents alike. State-of-the-art technology and careful waste minimization procedures will combine to allow the plant to operate without excessive impact on the environment.

INDUSTRY OVERVIEW: MIDVALE COMPUTERS

Working for You

Midvale Computers is among the fastest-growing companies in its field, rapidly opening new plants across the country. The plant proposed for your community will produce circuit boards for personal computers. Circuit boards are the backbone of the computer on which other parts, such as memory chips and microprocessors, are installed.

Making Circuit Boards

Circuit boards are made from rigid plastic sheets covered with a thin metal coating. Before the coating is applied, the plastic surface is cleaned with various liquids. These liquids dissolve dirt and other impurities, such as oil and grease, that do not dissolve in water.

The lines of metal that conduct electricity on the circuit boards are produced by a process called etching. Acid is used to etch (dissolve) large portions of the metal coating, leaving only ribbonlike lines of metal on the plastic. Once finished, the boards are cleaned again with solvent sprays; some parts must also be painted or coated with various materials.

Disposing of chemical wastes

Community Concerns

Community concerns about the operations of our plants are usually directed toward problems associated with the acid, metal, paint, and solvent by-products generated by the production of circuit boards.

Acid and Metal Wastes

The etching process produces an acidic waste with a high concentration of dissolved metals. These wastes cannot be disposed of without treatment. We neutralize the acid by adding measured amounts of base. We remove the metals by adding chemicals that combine with the metals to form solids. These solids are easily removed from the waste stream by filtering. We then discharge the treated wastes to the city sewer system.

The metal solids that are filtered out are treated further to recover valuable metals for reuse. We pump the remaining nonvaluable metals into a large, lined holding pond to settle. There, they form a very thick material called heavy metal sludge. This sludge is classified by the Environmental Protection Agency (EPA) as a hazardous waste because it can be dangerous to inhale or ingest, even in small amounts. According to federal law, this heavy metal sludge must be removed to a special landfill classified for hazardous wastes. The proposed plant will produce approximately two truckloads of the sludge per month.

Solvent containers are collected for recycling

Paint and Solvent Wastes

The cleaning solvents can be recycled and reused a number of times before disposal. Paint and solvent wastes that cannot be reused are also classified as hazardous by the EPA and must be disposed of safely. This plant will produce one truckload of paint and solvent wastes for disposal per month.

Groundwater Contamination

Recently, extensive groundwater contamination has been associated with some computer companies. In most cases, this contamination is a result of leaking underground storage tanks. All chemicals used at the Midvale plant will be stored in above-ground, double-walled tanks on paved floors and checked constantly for leaks. While accidental releases occasionally happen, these safety features help prevent chemicals from leaking into the ground, where they might contaminate the groundwater.

Worker Safety

Midvale Computers is recognized in many communities for its achievement in worker safety. We have one of the lowest accident rates in the field because of our strict safety procedures. These include safety training for new employees; use of safety eyewear, gloves, and other protective clothing; reduction in the number of employees who must come into contact with the chemicals; and a worker health-maintenance policy, which requires yearly physicals.

A Community Benefit

Midvale Computers is a successful company that produces quality products safely and with very little impact on the environment. No smokestacks, loud machinery, or other signs of an industrial presence will be noticeable from the roadway. In fact, the beautifully landscaped plant site will almost pass for an office complex or medical facility.

As is the case with any industrial development, the plant will place additional demands on the town's water supplies and sewage treatment system. Current predictions made by the company indicate that the town will have no trouble meeting its other needs while still providing for the needs of the plant.

Your community will benefit from the local taxes Midvale Computers will pay. Midvale means jobs and a healthy economy for your community.

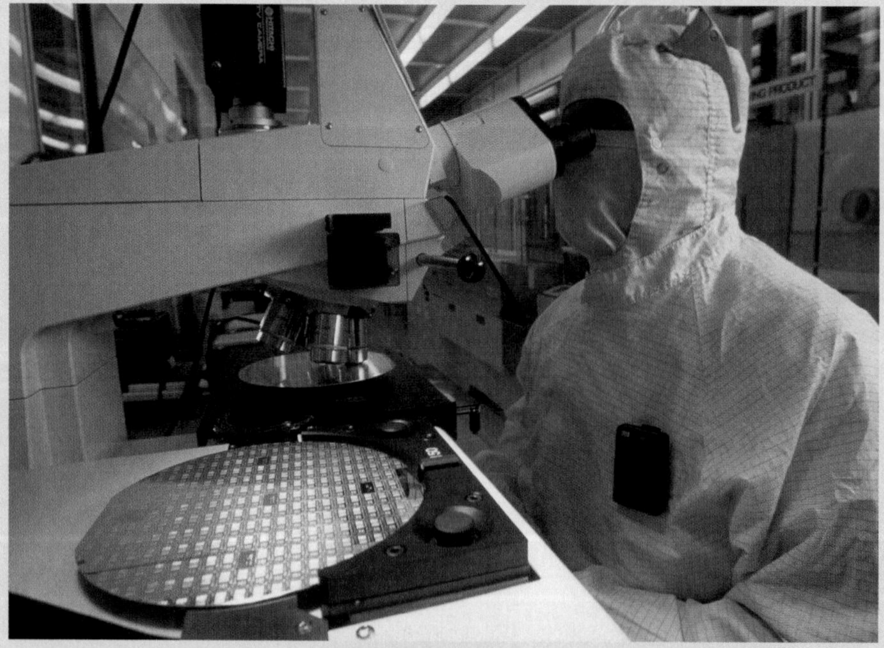

Examining chips for contamination in the "clean" room.

INDUSTRY OVERVIEW: RED RIPE FOODS

Working for You

The Red Ripe Foods Corporation operates a large number of factories across the country. Although the company produces a wide variety of fruit and vegetable products, it is best known for its tomato products—tomato paste, juice, sauce, and catsup. Our plans are to build a tomato-processing plant in your community.

From Field to Can to Grocery Store Shelf

After tomatoes are harvested, they are shipped in large trucks to the factory. We unload the tomatoes into water-filled storage tanks. Rollers move the tomatoes from the water bath through a series of sprayers. The rolling and spraying gently loosen surface dirt.

The cleaned tomatoes are then are trimmed and visually inspected. Next, they pass briefly through a steam bath, which causes the skin to loosen. Metal hoods above the steam bath release the steam harmlessly to the outside atmosphere so that it does not disperse throughout the plant, where it might be hazardous to plant workers. Pumps move the tomatoes to the pulper, which separates the skins and seeds from the tomato pulp. The seeds are collected, dried, and packaged to be sold; the skins must be discarded as waste. We pump the tomato pulp through glass pipes to be processed into juice, paste, sauce, or catsup.

Community Concerns

All industrial food processing requires attention to safety and the local environment. To limit the impact of plant smells and noise, we will site the proposed plant on the outskirts of your community. The trucks will load and unload at off-peak hours to minimize traffic congestion and the risk of accidents. As with many agricultural operations, work at the plant will be highly seasonal, depending on the availability of ripe tomatoes.

Cultures from factory equipment are checked for harmful bacteria.

Biological Hazards

We strictly monitor the chemical environment throughout every phase of production to control bacteria that can spoil food. As part of the canning or bottling process, we pasteurize all our products to kill microorganisms. The plant machinery is constantly supervised and cleaned frequently to minimize the possibility of bacterial growth. All wastewater passes through a fine mesh screen to remove tomato solids, which are sold as animal feed or as fertilizer. The small amount of tomato solids that cannot be reused must be disposed of as solid waste in the area landfill.

Wash water may contain some bacteria or other microorganisms washed from the tomatoes in the early cleaning stages. We treat this water, and all other water used at the plant, before release.

Catsup in your local store

Pesticide Contamination

We test all tomato solids, as well as all water used in the unloading and cleaning steps, for pesticide residues before disposal. Tomato growers must register the pesticides they use with local and state authorities, and they must provide this information—as well as dates of application—with each shipment to the plant. Red Ripe Foods does not accept produce from growers unless the legally required period of time has passed since pesticides were last used on that crop.

Community Benefit

Red Ripe Foods would like to open a plant in your community. The plant will provide employment opportunities for local residents, as well as food products that can be sold locally or shipped elsewhere. The management of Red Ripe Foods would like you to know about what goes on at the factory so that you can make an informed decision regarding the plant. Red Ripe Foods believes it can operate a plant that will provide jobs and needed products for the local economy with minimal environmental impact.

INDUSTRY OVERVIEW: STAR REFINERIES

Working for You

Star Refineries operates a number of facilities across the country. Its plants process crude oil to make many different kinds of products. Some of these products you may know and use. They include fuels for automobiles and aircraft; oils for lubricating cars; heating oils for homes; asphalts for roads; and the substances required by chemical companies to make many other products. These are used in fields as diverse as agriculture, sports, and medicine. Star Refineries wants the public to know about its operations so that residents of the community will be able to make informed choices about the plant location.

The Process of Refining Oil

The crude oil we process in our refineries was formed millions of years ago in underground pools or in certain kinds of rock formations. It is a mixture of different hydrocarbons, which are molecules that contain hydrogen and carbon.

After the oil is removed from the ground, it is transported to the refinery by pipeline, ship, truck, or some combination of the three. Pipelines will transport the oil to the proposed refinery. Once at a refinery, the first stop for the oil is usually large storage tanks where it is stored before processing.

Distillation towers break down crude oil into simpler compounds.

The first step in the processing begins at a distillation tower, which may be many stories high. Here, the crude oil is separated into the different kinds of hydrocarbons. This separation process works by boiling the crude oil. Lighter hydrocarbons with lower boiling temperatures rise to the top of the tower, and heavier hydrocarbons with higher boiling points are found at the bottom of the tower. Gasoline boils first, so it rises to the top in gaseous form. It is then drawn off and cooled, which causes it to become liquid again. Heavier hydrocarbons, such as those that make up heating fuels, collect in the middle of the tower, where they are removed. The heaviest components concentrate at the bottom of the tower, where we collect them for further processing into asphalts and the basic ingredients for the manufacture of a variety of chemical products.

After the crude oil is separated, each part must receive further treatment before use. About one-third of a barrel of crude oil processed by Star Refineries ends up as gasoline. The rest becomes jet fuels, kerosene, lubricants, thinners, solvents, and many other products.

Community Concerns

Communities in which refineries are located are generally concerned about the visual impact of the facility, the risks of fire and explosion, the effects of oil leaks or spills on the environment, and the effects of the refining process on air quality. Star is an industry leader in the development of procedures to prevent accidents and minimize environmental impact.

Company Headquarters, Clark County

Visual Environment

The plant will be located in the existing industrial area near the river north of town, far from any residential areas. Star Refineries will construct a parklike border of trees around the plant and along the opposite shore of the river. This will help absorb noise and reduce the visual impact of the refinery.

Fires and Explosions

Our procedures to prevent fire and explosion include constant checking of storage tanks and plant equipment for any signs of leaks or other problems. To minimize the risk of fire, the large storage tanks that hold oil prior to processing are located a great distance away from other plant equipment.

Air Pollution

Techniques for preventing the release of airborne pollutants have improved greatly. We collect the hydrocarbon vapors and burn them off cleanly, pro-

Filters trap hydrocarbon particles.

ducing carbon dioxide gas, water, and a little smoke. The carbon dioxide gas has the same chemical composition as the carbon dioxide in your breath and is nontoxic in the concentrations produced by the plant. The distillation towers contain filtering devices that remove particles from the refinery exhaust. These hydrocarbon-containing wastes are either recycled or treated. One treatment method is to use oil-consuming bacteria that break down hydrocarbons to release carbon dioxide gas and water and create organic material that can be packaged and sold as a soil additive.

Water Pollution

The refining process requires millions of gallons of water daily, much of it to keep the towers cool. We recycle most of the water and reuse or treat it to render it nontoxic before releasing it to the municipal wastewater system.

A Community Benefit

Star Refineries would like to build a plant in your community. The management of Star works closely with government regulators to maintain its excellent record and its reputation as an industry leader in safety and environmental issues. Star developed many of the standard industry procedures for reducing dangerous emissions and recovering wastes. Star Refineries will bring many good, safe jobs to your community.

ANALYSIS

1. What subjects did the industry statement cover well, and which ones did it not seem to cover in depth?
2. Share your results with the other members of your group.

2 Pinniped Island

Pinniped Island and Its Future

Imagine you live on Pinniped Island and are concerned about its future. The lack of good, year-round jobs causes many young people to leave the island. Your interest in the island's future makes you want to know more about the island and its resources.

CHALLENGE

Read about Pinniped Island and its current resources. Think about the advantages and disadvantages of bringing new industry to the island.

Legend

Park
Forest
City or town
Main road
Ferry lines
0 1/2 1 Kilometers

Map of Pinniped Island

Bass River
High Mountain Lakes
Shepherd's Point
Rockville
Fishhead Bay
Fishhead Point
High Mountain Park
Wright's Bay
Wright's Ferry
Clamshell Bay
Clear Lake
Clamshell Beach
Gull Cove
Clear Lake Resort
Seal Harbor
Buck's Harbor
Point Joseph
Clear Lake Park
Gull Cove
Klickitat
South Beach

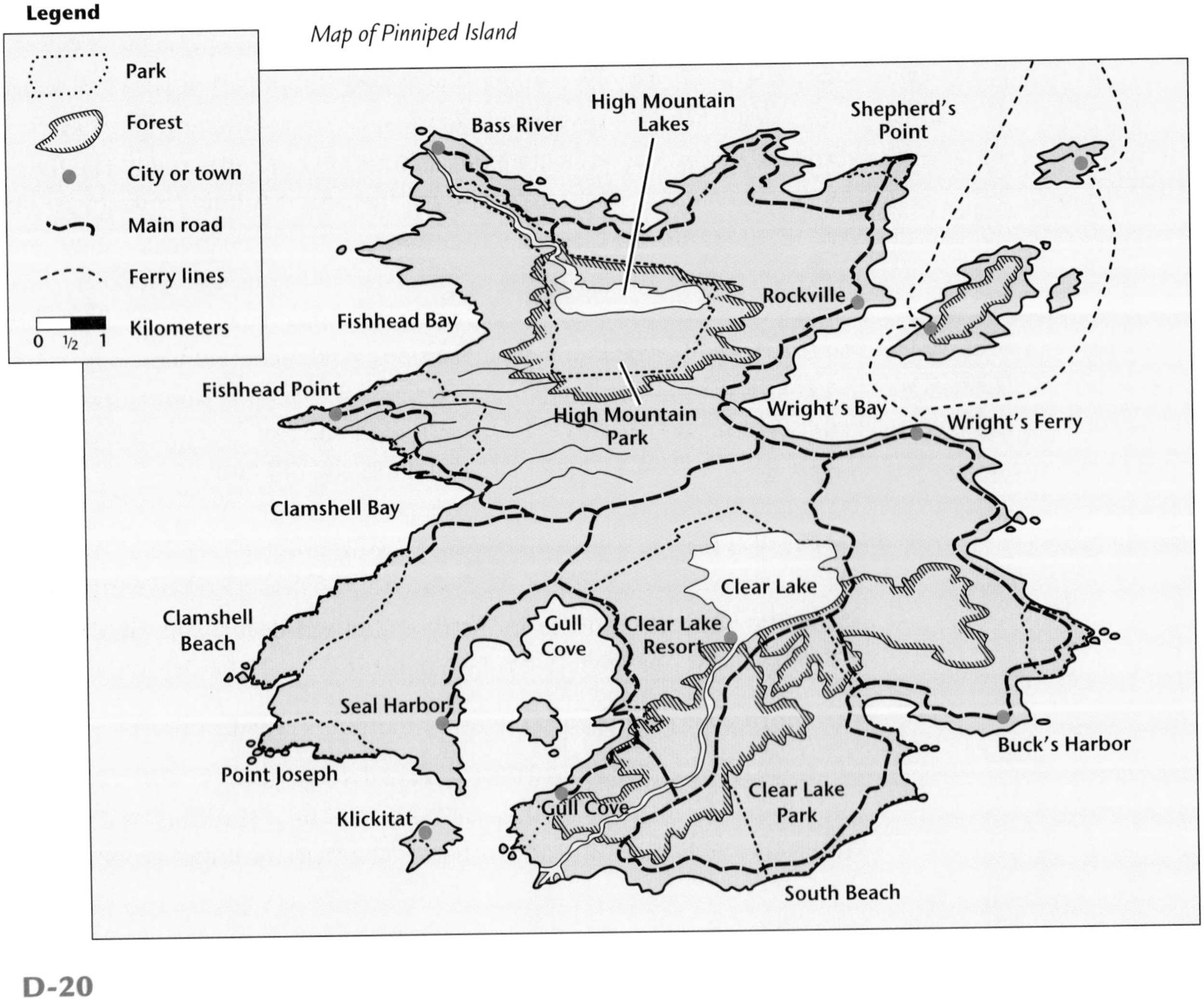

High Mountain Lake

Pinniped Island is located about 30 miles off the coast. Irregular in shape and fairly mountainous, it is roughly 20 miles long and 15 miles wide. The highest mountains are in the northwestern part of the island. Mount Reyes, in the High Mountain Park area, is the highest peak—5,253 feet. The eastern side of the island is hilly, with no peaks above 3,000 feet.

Much of the island's yearly rainfall of 25–30 inches is captured in the island's many small streams and lakes. Clear Lake, which is located in the south central region, serves as the island's main reservoir. High Mountain Lake is a smaller lake in the northwestern region of the island that is isolated from the main river system.

The island's climate is mild year-round with average daytime temperatures between 5 and 10°C (approximately 40 and 50°F) in winter and 20 and 30°C (approximately 68 and 87°F) in summer. Winds are generally from the northwest and can be quite brisk on the western side of the island.

Sheep and cows graze the island's extensive grasslands. Forested areas are limited, and large trees are sparse except near the lakes and rivers. The island does not have any minerals of economic importance.

Wright's Ferry

Many of the island's 15,000 people live in small towns along the coast. Wright's Ferry (population 4,200) and Buck's Harbor (population 2,900) are the largest towns. Tourism, fishing, raising sheep for wool, and dairy farming are important to the local economy. Travel to the mainland is by ferry; the express takes approximately 1 hour, while the ferry that stops at neighboring islands takes 1 hour and 45 minutes.

ISLAND CLARION

Last night the Island Council discussed locating a glue factory on the island. Some members of the council think a glue factory would improve the economic outlook on the island. Today the following comments were printed in the paper in answer to the question: Are you in favor of the factory?

CHALLENGE

Read each response carefully. For each response decide what part is based on evidence and what part is based on personal opinion. What other evidence would you like to have to help you make a decision about the factory?

Pinniped Island Clarion

June 3, 2002

What's your opinion? Are you in favor of building a glue factory on Pinniped Island?

Jacob Smith, self-employed concrete contractor
"Sure, I'm all for it. There's not enough work on this island, and a glue factory will mean a lot of work building roads and factory buildings. Unemployment on the island is nearly 20% and even higher in the winter. I heard that the factory will provide nearly 200 jobs. This is an important step in improving our economy."

Cameron Davis, shipping clerk
"Think of all the jobs it will create, not only in the factory, but in services needed to support it. Our shipping and cargo company will certainly increase the number of shipments that are transported both to and from the island."

Melinda Phipps, school guidance counselor
"Yes. I'm tired of seeing students from our high school leave to get jobs on the mainland. It's better for them to work here and remain part of the island community."

Alana Teagues, computer programmer
"No. The factory will destroy the environment and only help the fat cats who operate it. All factories are dirty and noisy, and you know how they create pollution."

Isaac Roberts, middle school student
"Maybe. I don't know. It depends on a bunch of things. It's difficult to decide without knowing more."

John DiSarro, high school senior
"Yes, because then I can get a job and pay for my car insurance."

Hans Efger, fisherman
"No. The fish populations are decreasing, and the waste from a factory would destroy the fishing industry on this island."

Raylene O'Malley, Island Utility Department
"No. Our energy analysis shows that we have enough power to meet current needs, but a factory will use a large part of our reserves. We will need to build another power plant with money we don't have if the island continues to grow and needs more power."

June Chan, neighbor of existing dairy
"No. The factory will need more milk if it's to grow and make money. The dairy I live next to is a source of flies and foul smells. There will be more trucks coming and going to ruin our rural atmosphere."

Fernando Castillo, proprietor, The Island Cafe
"I don't know yet. It's true that our unemployment rate is a problem. I would be concerned that a factory would cause pollution and traffic problems. I'm waiting to find out more before I make up my mind."

Producing White Glue

In order to evaluate the impact of a glue factory on the island, you will need to gather evidence about the products and by-products of glue production. When milk is warmed or acidified, most of its protein content precipitates out as *curds*. What is left is a solution containing milk sugars and water soluble proteins called *whey.* The curds from curdled milk, called *casein,* produce a form of glue that is very strong and water resistant.

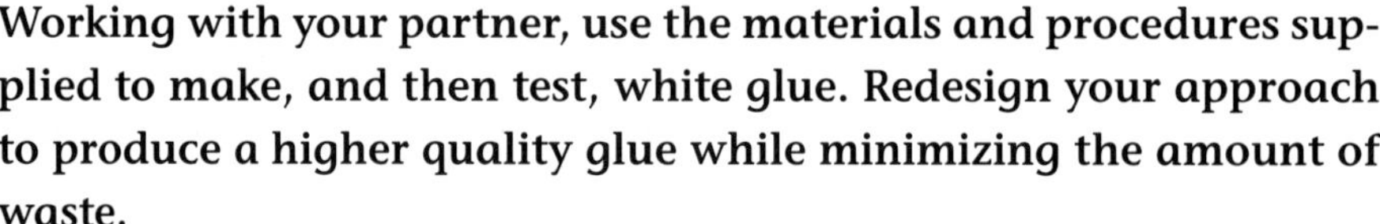

Working with your partner, use the materials and procedures supplied to make, and then test, white glue. Redesign your approach to produce a higher quality glue while minimizing the amount of waste.

MATERIALS

For each group of four students

1 30-mL dropper bottle of 5% household vinegar
1 30-mL dropper bottle of 5% household ammonia
pH paper or universal indicator solution
materials to glue together, such as small pieces of paper, wood, plastic, aluminum foil, glass slides

For each pair of students:

1 SEPUP tray
1 SEPUP filter funnel assembly
1 graduated container
1 10 mL of warm (40°–50°C) milk
1 piece of filter paper
1 stir stick
1 paper towel
1 dropper
containers for storing glue

PROCEDURE

Part One: Making White Glue

SAFETY NOTE: Be sure to wear safety eyewear during this investigation. Take care when handling hot liquids.

1. Before you begin making the glue, plan the two tests you will use to determine the quality of your glue.
2. Add 10 mL of warm milk to the graduated container.
3. Add 12 drops of vinegar to the 10 mL of warm milk. Stir after the addition of each drop.
4. Filter the mixture, separating the curds (the solids) from the whey (the liquid). Put the whey in the waste container.
5. Transfer the curds from the filter paper to the graduated container and rinse them with clean water.
6. Mix the curds with 12 drops of ammonia. Warm the mixture, as necessary, by putting the cup in a container of warm water and stirring until all the curds are dissolved.
7. If you think the glue should be thinner, carefully add more water, a drop at a time.
8. Perform two tests to determine the quality of your glue. Label the glued objects and place them in a safe location until the next class session.
9. Clean all your materials thoroughly to remove any traces of glue.

ANALYSIS

In your science notebook, record your experimental procedures for making and testing the glue and your observations. You will share your results with the class tomorrow.

Part Two: Testing and Redesigning the Glue

1. Observe the tests you set up in the previous session.
2. Make a table to record your test results. For each test, you should record the glue characteristic you are testing, the procedure you used, your test results, and conclusions.
3. Discuss your results with your group or your class.
4. Based on the discussion, work with your partner to plan the changes you will make to improve your glue recipe.

ANALYSIS

Record your plan for making an improved glue. Be sure to describe why you think your plan will result in better glue.

Sample 1: Tests of Glue Characteristics

Glue Characteristics Tested	Procedure	Results	Conclusions

PROCEDURE

Part Three: Making a New and Improved Glue

1. Follow your plan to make your new batch of glue.
2. Use the same tests you used before to test the new glue.

ANALYSIS

1. Record the procedure you used to make an improved glue.
2. Compare your new glue to the first batch you made. How did you change your glue-making procedure? Did the new procedure make an improved glue?
3. How would you modify your process and/or ingredients to help you maximize the production of white glue while minimizing the amount of wastes produced?
4. Discuss the changes you made with the other team in your group. What were the similarities and differences in the redesigned approach that each team adopted for making a better glue? Come to a consensus on a process your group will use for scaling up production in the next activity.

4 Scaling Up Glue Production

How Much Does This Island Need?

You have produced a small amount of glue. Now you want to know if there is a market for the product. You will begin by scaling up the production of white glue from experimental lab amounts to a quantity large enough to meet the needs of the test market. Based on your findings, you will develop a plan to scale up your process so that you can make enough glue to supply everyone on the island for a year.

CHALLENGE

Design a process for scaling up your method for making glue.

Imagine scaling up cookie production from a few dozen homemade cookies to thousands of cookies for a national market.

PROCEDURE

1. Working with your group, use the island information tables on the next page to choose a proposed test market for your glue.
2. Decide how many 60-mL (2-ounce) bottles of glue you will need for test marketing.
3. After discussing the various proposals with other groups in the class, record in your science notebook the size and makeup of the agreed-upon test group and the number of bottles of glue you need for test marketing.
4. Working with the other members of your group, design a plan for scaling up to produce the quantity of glue needed for the test market. Make sure that your plan includes at least:
    - the amount of each ingredient you will use;
    - the approaches to production you will use; and
    - how you will minimize and then handle the wastes produced.
5. Be prepared to present your plan to the class.
6. Based on the reports and discussions about scaling up for test marketing, and the island information given on the next page, estimate the amount of glue that would be needed to supply Pinniped Island for one year.
7. Use your estimated market for the whole island to design a plan for making enough glue for the island for one year.

Making a glue test product

ANALYSIS

Possible factory site, north side of island

Produce a report that describes how you made your estimate of the size of the glue market on Pinniped Island *and* how you used that estimate to scale up the glue recipe. Be sure to show all the steps of your math work. (Other people might not agree on your "market estimate," but don't worry about that. Do the best you can, and *explain your thinking* so others can react to it. Then use the number you came up with to scale up the glue recipe.)

ISLAND INFORMATION

Population	
Children (newborn—20)	5,000
Adults (21—55)	8,000
Seniors (over 55)	2,000
Total	**15,000**

Residences	
Residences with children	2,700
Residences without children	2,300
Seniors (over 55)	2,000
Total	**7,000**

Schools	
Day care centers	4
Preschools	2
Elementary schools	5
Middle schools	2
High schools	1
Colleges and trade schools	2
Total	**16**

Packaging the Glue

You will compare your group's glue package to the commercial one.

CHALLENGE

Rate and compare the two packages based on a ten-point scale for each characteristic considered important by the class.

PROCEDURE

1. Produce a table like the one on the next page for your group. In the extra spaces at the bottom of the table, include all other packaging characteristics the class considers important.
2. Using a ten-point scale for each category, rate your group's package and the commercial one.
3. Record the total score for each package in your science notebook.
4. Before discussing your answers with the other members of your group, answer (in your science notebook) the three questions listed below the table.
5. Discuss your answers with your group and come up with a consensus.

Comparing Your Package to a Commercial Package

Categories	Your Package	Commercial Package
Cost		
Strength		
Can it be reused or recycled?		
Environmental impact as waste		
Convenience		
Labeling		
Total Points		

QUESTIONS

1. Based on your results above, what kinds of changes could you make to your glue container to help improve its overall rating?
2. What kind of trade-offs did the manufacturer make in producing the commercial glue container?
3. Why do you think it is possible or impossible to produce the perfect (ideal) glue container?

Plan Your Factory

The Pinniped Island Council has agreed to consider proposals for construction of a glue factory on the island. Your company must submit its factory plan and a complete environmental impact report. All submissions will be evaluated competitively by the council to decide whether one of the companies' proposals to build a glue factory will be approved.

CHALLENGE

Think of your group as a company and design a factory and total plan for producing glue on Pinniped Island. Prepare an environmental impact report that justifies your company's design and siting decisions.

A new glue factory might lead to an increase in local traffic on the island.

PROCEDURE

Part One: The Preliminary Plan

1. Use the Factory Planning Guide to help you begin your plan.
2. Develop a company (group) statement of your preliminary ideas about how to meet each of the agreed-upon needs for a factory on the island. Reflect on what you have learned throughout the year. Use your science notebook and this student book as resources to help you plan.
3. You may want to identify several alternatives to choose among later. Your company's statement should help you identify additional information that you will need to make final decisions and to prepare the environmental impact report.
4. Include rough maps and/or sketches to illustrate your project.

FACTORY PLANNING GUIDE

1. **New construction**
   Glue production will require factory buildings. Your plan may also call for additional roads, docks, or other construction for transporting the raw materials and the finished product. Describe everything that will be built as part of your project.

2. **Material resources: Milk**
   You will need a source of milk. There are three existing dairy farms on the island. A typical farm there covers nearly 28 hectares (70 acres) of land. Each farm has about 500–600 cows on the dairy, or 30–38 cows per hectare (12–15 cows per acre) of pastureland. The number of cattle is restricted to help preserve the pasture and the groundwater below it. At any given time, there are 500 cows involved in daily milk production. The dairy produces 20,280 liters (5,500 gallons) of milk per day, or 41 liters (11 gallons) on

the average from each cow. The dairy farms use their own well water to water and wash the cows, as well as for milk production and irrigation. Most of the cow feed comes from grain grown on other farms on the island. The rest is from the farms' pastures.

Will the existing dairy farms meet your needs, or will you need to increase dairy farming or import milk? How will either of these choices impact the island? Consider your information from the scaling-up activity and plan how you will obtain enough milk.

3. **Material resources: Other materials needed**
   You will also need sources of vinegar, ammonia, and other supplies. Consider the properties of these materials as you decide how you will obtain, transport, and store them.

4. **Human resources**
   Workers are needed to run the plant. How many workers do you expect to need? What kind of skills will they need for the kinds of jobs you will have? Where will they live? Are there enough people on the island to fill the jobs, or will you have to hire people from the mainland? The production process may be altered so that less labor is required to operate the factory, but this change would double the amount of energy required to run the plant. More oil would have to be imported or more water power used, which would drive up the energy costs.

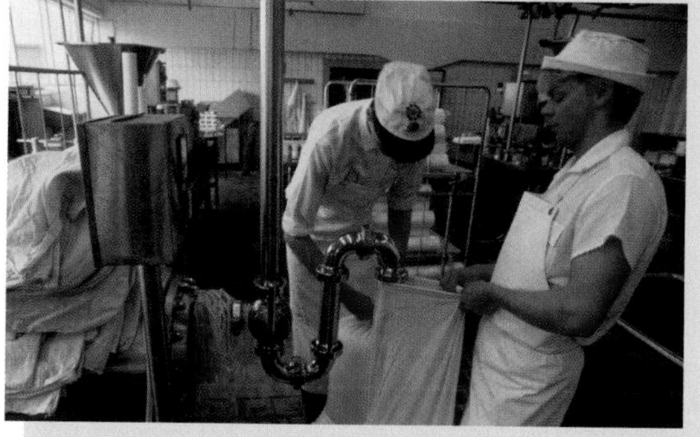

5. **Energy resources**
   You will need energy for power to run the factory machinery, lights, heating system, and other tasks. You must consider whether the current energy supply can meet your needs. If not, you will have to find a way to obtain energy by expanding existing power plants or finding new energy sources. You must consider the environmental impact of any choice. Oil and gasoline are needed to run cars and trucks and to lubricate

machinery. To import oil, you will need to have a transport system—ships, docks, pipelines, trucks—in place. Solar power can meet some of your energy needs but, except for heating water for the plant, solar power is a more expensive and seasonal energy alternative.

6. **Water**
The island currently has enough water to meet its needs. The western and central areas use groundwater. In the southeast, most people have their own wells, or they take water from streams or small reservoirs. In planning your factory, you must consider the quantity and quality of the island's water supply.

7. **Land use**
A limited number of sites on the island are suitable for construction of the factory and related facilities. You must take care to come up with a plan that provides for safety and has the least negative effect on the island's environment. The tourism industry is very concerned about which land is used for the factory. The location of the factory will also affect the costs of transportation for workers and materials. Some land use can be decreased by providing for certain needs off the island.

8. **Waste disposal**
How can you treat and/or dispose of factory wastes? Consider the impact of your waste treatment plan on the environment and on human health and safety.

PROCEDURE

Part Two: Environmental Impact Report

Your environmental impact report should be a balanced presentation of both the negative and positive impacts of the factory on the natural and human environment of Pinniped Island. Your company will have 10 minutes to present its plan to the council. Your plan will be rated according to how well it considers community concerns. In preparing your report:

- Give careful consideration to the background information provided to the Town Council on pages 40–47.
- You are encouraged to add additional information to your report to strengthen your presentation.
- Cover at least the seven items listed below.

1. **Factory proposal**
   Describe your factory in detail. Consider the entire life cycle of the glue, from obtaining the materials needed for packaging and shipping the product to its final market. Choose and justify the site of your factory. Your plan should describe the proposed physical layout of the factory and the chemical interactions that the factory will use. If you plan to build power plants, roads, docks, or other facilities on the island, you should indicate the locations of these facilities.

2. **Existing environment**
   Where will you locate the factory? Describe the environment as it now exists.

3. **Factory impact on the natural environment**
   Describe how your project will impact the natural environment. Include both positive and negative effects. Be sure to include all

of the following aspects of the natural environment as well as any additional topics you think are applicable.

a. surface water

b. groundwater

c. vegetation

d. animal life

4. **Factory impact on the human environment**
Describe how your factory will impact the human environment. The "human environment" refers to any feature of the environment related to human activity, from buildings and cities to economic, social, and political systems. Your report should include the following areas as well as any additional areas that you think are important.

   a. water supply

   b. transportation systems

   c. waste systems

   d. community organizations (such as schools, recreational facilities, housing, hospitals)

   e. economic factors

   f. quality of life (such as noise, lighting, air, visual environment, factors that could involve health risks)

5. **Mitigating impacts**
Describe how you plan to mitigate (lessen or compensate for) any negative impacts you described in Sections 3 and 4. Also describe any impacts that cannot be mitigated.

6. **Summary statement**
Write a brief statement describing the major trade-offs that you encountered in developing your plan. Indicate why you think that your plan is the best overall factory plan possible.

7. **Resources**
Describe the resources that you used in preparing your report.

Water Report

Prepared by Pinniped Island Water and Sanitation Company

Clear Lake and Wells 1 and 2, currently owned by the company, provide the drinking water supply on Pinniped Island. We removed Well 3 from service in 1993, when iron levels increased with a rise in the acidity of the water. We are investigating recent increases in nitrates in Well 1, although they are still well below federal standards. Clear Lake serves as a reservoir for emergency water needs and is used for recreational purposes, including swimming, boating, and fishing. Because of drainage into the lake from unknown sources, the level of THMs has increased by 0.01 ppm over the last year.

Currently, Clear Lake and Wells 1 and 2 are able to supply 180 gallons of water per minute to the water treatment plant in Wright's Ferry. Here, the water is chlorinated and filtered before it is pumped to customers in the service area. The company has 2,320 metered customers, and serves more than 8,500 people. The average person uses 27 gallons of water per day on the island. Other people on the island obtain their water from streams, small reservoirs, and private wells.

Your water company currently operates a wastewater treatment plant located to the south of Wright's Ferry. The plant treats wastewater (effluent) to a "primary level." The effluent is first placed in large settling tanks to separate out liquids and solids. The remaining liquid effluent is chlorinated and pumped through an underwater pipe one-half mile out into the ocean, where it disperses in deeper water. The solid material is composted, dried, and sold to the island residents for use as soil amendments. The water treatment system handles 175,000 gallons of waste-water per day and has a capacity of 200,000 gallons per day.

Water Report (cont.)

Federal Water Quality Standards and Pinniped Island Water Sources

	Federal Standards (ppm)	Clear Lake	Well 1	Well 2	Well 3
Chloride	250.0	47.0	15.5	14.0	220.0
pH	6.5–8.5	6.6	7.2	6.8	6.3
Turbidity	1.0 (NTU)	0.1	0.2	0.2	1.3
Coliform bacteria	<5%	3.5%	1%	0.5%	0%
Total dissolved solids	500.0	120.0	87.0	85.6	520.7
Nitrate	10.0	1.4	3.2	0.6	20.1
Iron	0.30	0.10	0.25	0.25	0.60
Trihalomethanes (THMs)	0.10	0.02	0.03	0.02	0.01

Many people on the island use septic tank systems, where their household wastewater is collected in large underground tanks. Here, solids settle and bacteria help decompose the mixture. The liquid portion in each tank overflows into a pipe that leads to a group of branching, porous pipes buried underground—the leach field. In the leach field, liquid wastes feed into the subsoil and gradually spread out away from the home.

Energy Summary

PINNIPED ISLAND ENERGY FUTURE—POWER FOR THE 21ST CENTURY

Prepared by SR Associates for the Pinniped Island Business Council

Energy generation

Pinniped Island uses one fuel-oil-powered generating plant located on the coast, southwest of Pinniped Ferry. Along with fuel oil, gasoline, propane, and diesel fuel are imported and stored in a "tank farm" near the power plant. Island Fuel and Service Company distributes these fuels to businesses and homes across the island.

Current energy use

The pie chart shows Pinniped Island electrical power usage. The graph on the next page shows the historic demand for power.

The generating facility is capable of producing up to 500,000 kilowatt hours of electrical energy per day. About 14,500 customers currently use the system.

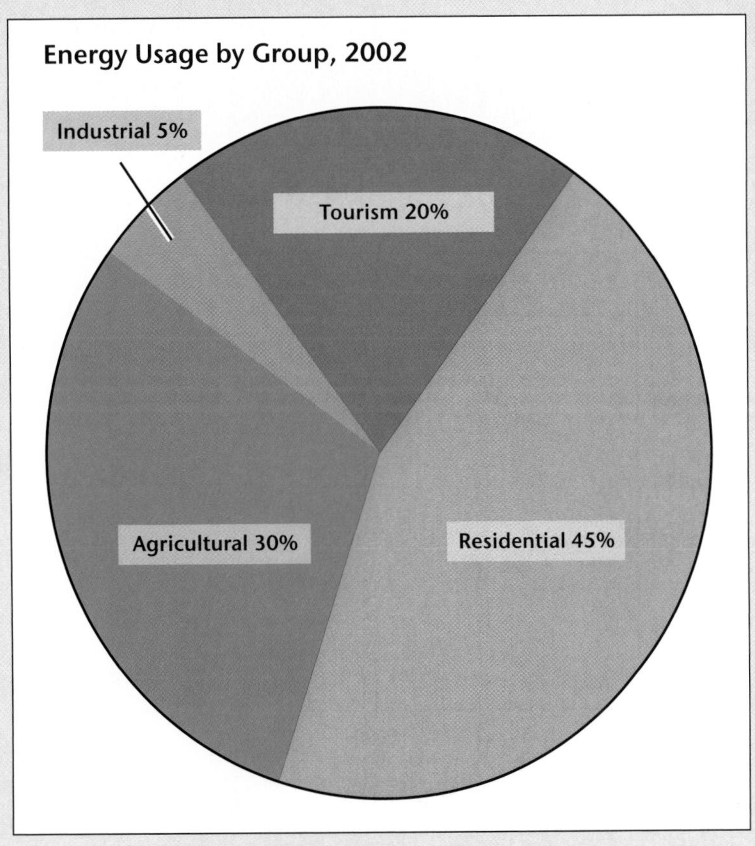

Energy Summary (cont.)

Future outlook

With increasing fuel prices, the island must consider other alternatives to its present dependence on imported petroleum products. Several alternatives are available: strict conservation measures to reduce electrical demand; the construction of a wind farm at the northern point of the island; the use of solar collectors and cells to generate energy, using large amounts of land to generate solar photovoltaic power (the island has over 200 sunny days per year); and the conversion of biomass from sheep and cattle. The latter would create methane gas that could be burned in the electrical plant to create energy.

If these alternatives are not acceptable, the only way to increase the amount of power generated would be to add a second fossil fuel-fired generating facility. This facility would cost $17 million to construct and would require more imported fuels. Another alternative would be to dam some of the island's rivers to create hydroelectric plants.

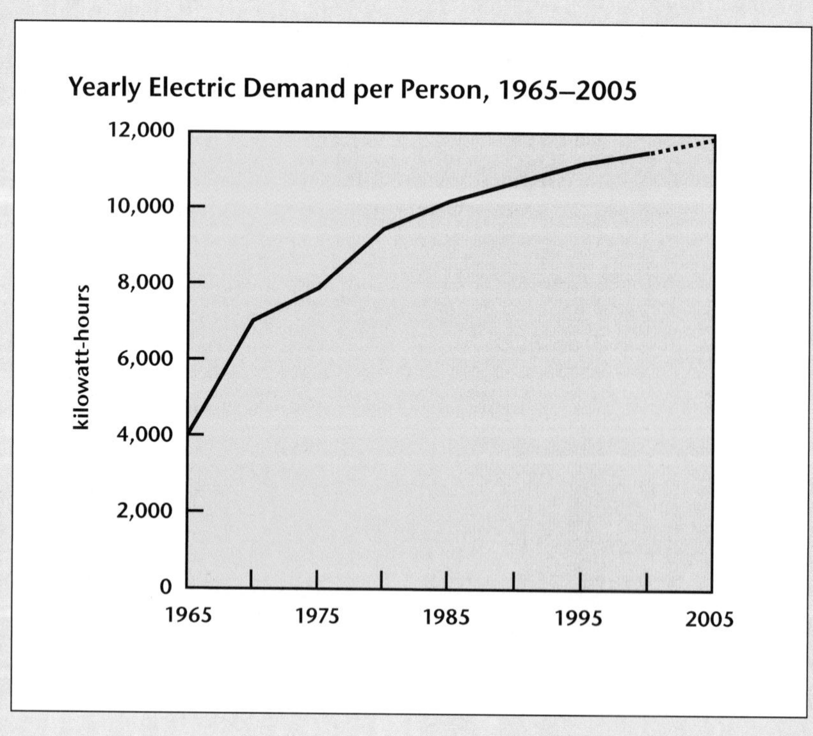

Land Use Report

Prepared by Pinniped Island Bureau of Land Management

The Pinniped Island Bureau of Land Management manages the island's parks and recreation areas and forests. The BLM also works with island residents to study agricultural land use and other uses of island resources.

Public Lands

The BLM manages the island's two regional parks (High Mountain and Clear Lake Parks) and numerous smaller recreation areas. Our newest area is the Buck's Marsh Wildlife Refuge north of Buck's Harbor. This refuge has been set aside to protect the many species of birds and other wildlife on the eastern shore. Populations of birds and other wildlife remain healthy on the island, and we hope to continue our tradition of respect for the island's diverse wildlife.

Plans are under way to protect parts of the forest east of Clear Lake in the future. Other forested areas are important in timber production, which is restricted to meeting island needs. Lumber is not exported from Pinniped Island.

Mining resources on the island are limited. The limestone and marble quarries in the western areas are controlled by the BLM.

Nearly 70% of the agricultural land on Pinniped Island is made up of large ranches and farms that are devoted to grazing sheep and cows or raising corn and wheat for livestock feed. The remaining agricultural land consists of a number of small farms that primarily produce food crops.

In planning for future agricultural land use on Pinniped Island, it is important to note that a substantial portion of the island (20%) is extremely hilly or mountainous and not suitable for agriculture. The central valley, parts of the area south of Wright's Ferry, and areas of the northern and southwestern peninsulas have suitable land that is not yet devoted to agricultural use. Development of these areas as farmland should be accompanied by careful studies of the impact of farm animals on vegetation and water systems.

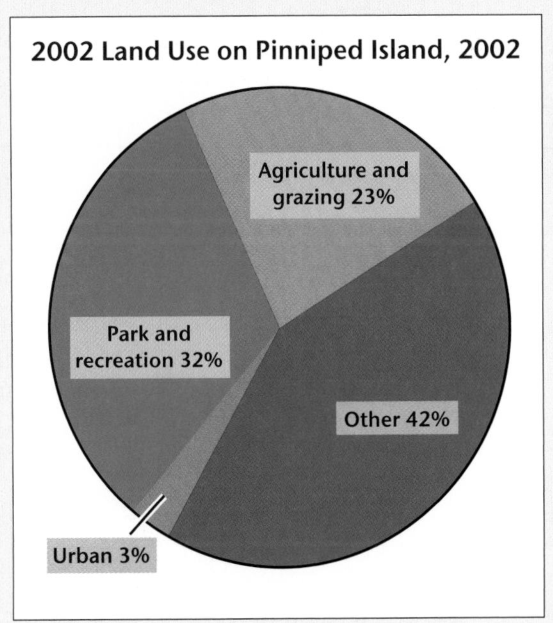

Employment Report

Pinniped Island Department of Vital Statistics

The island's year-round population increased rapidly in the 60s and 70s. Since 1980, the population has grown more slowly. In 2000, the unemployment rate hit an all-time high, and has remained above 15%.

Job	Percentage of Population	Median Income
Fishing	15%	$30,000
Tourism	32%	$38,000
Dairy farming	4%	$36,000
Sheep raising	4%	$35,000
Other farming	4%	$31,000
Professionals*	7%	$53,000
Labor**	4%	$35,000
Small businesses***	12%	$31,500
Other	8%	$20,000
Unemployed	10%	unknown

* Includes doctors, lawyers, teachers, engineers, and other positions generally requiring a college degree.

** Includes construction workers, transportation workers, etc.

*** Includes small businesses that serve year-round and summer residents.

Employment Report (cont.)

Year-round Residents

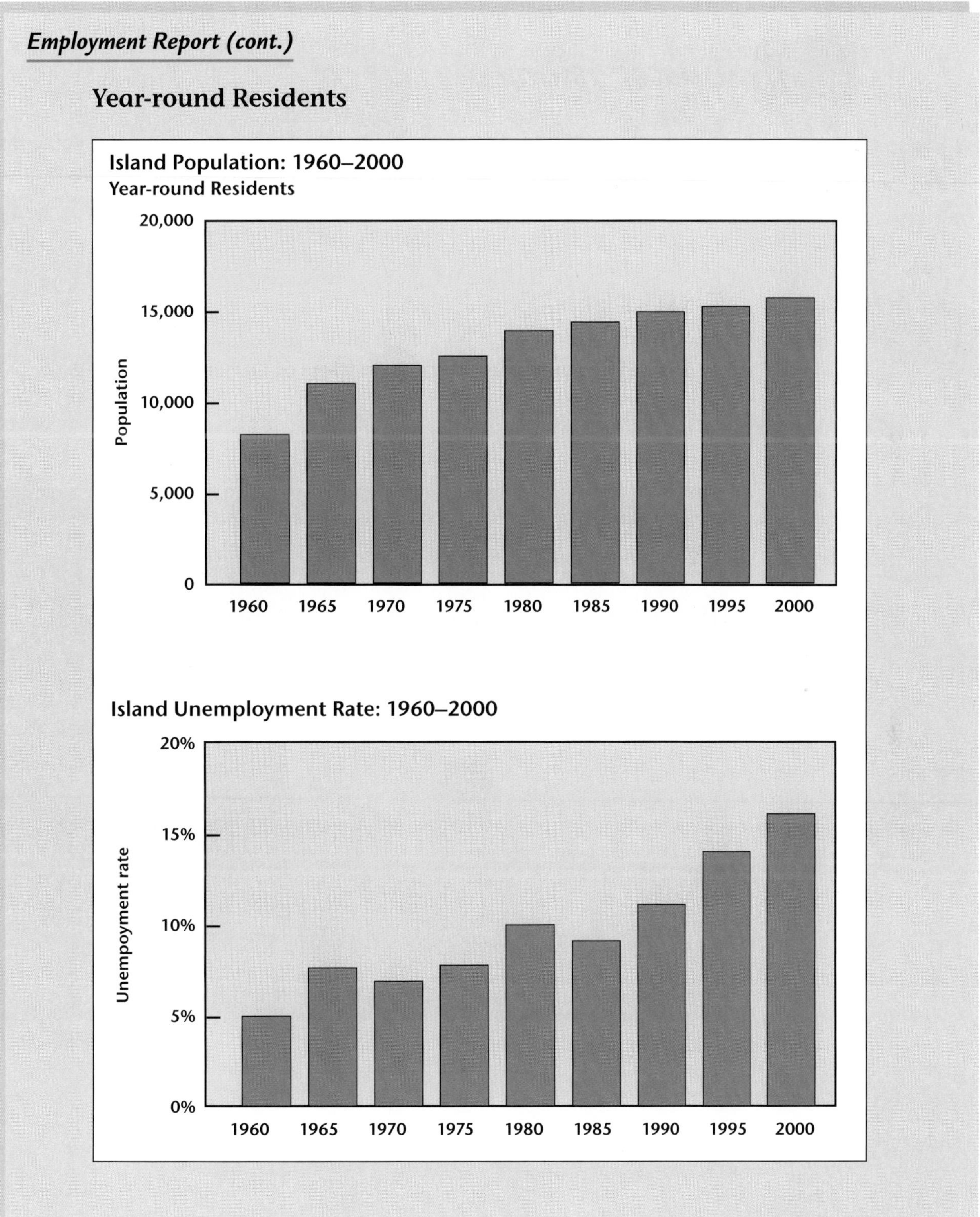

7 An Island Parable

Easter Island

The story of Easter Island describes what can happen when people do not consider all the environmental implications of their actions.

CHALLENGE

Identify the problems that the settlers of Easter Island faced.

Easter Island statues

TURNING ABUNDANCE TO SCARCITY

They came in long canoes, riding majestically across the clear blue waters of the Pacific. They came, bringing hundreds of vegetable and fruit plants to cultivate for food.

They came to a wind-swept island full of forests and flowers, covered everywhere with the volcanic debris of the towering brown mountain.

Their plants grew and they prospered. Fields of bananas, yams, and ti (a tropical palmlike tree) covered the rich brown soil. They cut trees to make more canoes to fish the abundant waters and build their homes. Soon, their population grew, and they needed more land to grow food and more wood to shelter their growing numbers against the fierce, damp winds of winter.

Out of the volcanic rocks of the mountain they carved huge humanlike statues, which they hauled down its slopes on wooden sleds to be erected as giant figures on platforms by the edge of the sea, staring out into the vastness of a never-ending ocean.

Newcomers—The Crowning Blow

Over many hundreds of years, the forests were slowly cut down. And as this happened, the newcomers arrived—the ones with "big ears" who came in their canoes. They, too, desired to share the bounty of the island. But they scorned the old ways of the islanders. "Why spend time carving stones when there is so little food to be had here?" they asked.

It was true. The once-rich island was now treeless and full of houses—people even lived in rocky caves formed from long lava tubes. The people now numbered nearly ten thousand. There was no longer time to create those majestic statues. Now there was only time to survive and guard one's food and shelter from the newcomers.

The newcomers fought for land and eventually overcame the statue-carving "short ears," whom they had grown to hate. In a final battle, nearly all of the short ears were obliterated from the island. This one-day war ended with all of their bodies being placed in a giant fire in long trenches dug along the side of the dead volcano.

But soon, too soon, the food supply ran short. Eventually, the big ears turned to cannibalism. Skeletons filled the empty caves. The remaining wood was burned to cook food and fend off the bone-chilling winds of winter.

One day, a ship arrived to find fewer than 50 natives, eking out a meager existence on a rock-filled island covered with short grasses and the remains of village dwellings. And yes, there were those mysterious statues, made in a far distant time, looking out to the vast sea.

The name of the island, you ask? It is called Easter Island, found 2,300 miles off the coast of Peru and nearly 1,200 miles from the nearest island in Polynesia.

QUESTIONS

1. What could the settlers of Easter Island have done to prevent the problems they faced?
2. What does this story tell us about our own relationship to our environment?

Index

C

I

K

N

O

P

T

Y

Illustration and Photo Credits

Abbreviations: t (top), m (middle), b (bottom), l (left), r (right)

Cover: all photos ©PhotoDisc except for water mill: Michael Townsend/Stone

All illustrations Seventeenth Street Studios except as noted.

Water

Activity photo/icon: ©PhotoDisc; Unit opener: A-1 ©Corbis; Part 1 opener: A-2 tl: ©PhotoDisc, tr above: ©PhotoDisc, tr below: Susan Spann, ml: ©PhotoDisc, m: Gelderblom/Eye of Science/Photo Researchers Inc., mr: Frank Siteman/Stock Boston, bl: ©PhotoDisc, br: Daniel Brody/Stock Boston; A-3 ©PhotoDisc; A-4 Donna Markey; A-9 Susan Spann; A-11 Susan Spann; A-15 Lab-Aids; A-16 Mike Reeske; A-21 Herbert D. Thier; A-25 Map of London, adapted from Map 1 (prepared by C.F. Chiffins, Lith.) in John Snow, *On the Mode of Communication of Cholera,* 1855; A-26 Bettmann/Corbis; A-27 Photograph, 1856 or 1857, published by Benjamin Ward Richardson in 'John Snow M.D.', *The asclepiad,* 1887, vol. 4, facing page 274. The Wellcome Library, London; A-28 courtesy Mary Evans Picture Library; A-29 Bettmann/Corbis; A-33 Lawrence Migdale/Stock Boston; A-35 Jerome Wexler/Photo Researchers, Inc.; A-36 Bob Daemmrich/Stock Boston; A-40 Susan Spann; A-41 Susan Spann; A-42 Lab-Aids; A-44 t: Lab-Aids; A-48 Susan Spann; A-49 CNRI/Science Photo Library/Photo Researchers Inc.; A-50 Frank Siteman/Stock Boston; A-51 Lab-Aids; A-53 Joe Sohm/ Stock Boston; A-56 Susan Spann; A-57 Susan Spann; A-63 ©2002 Timothy O'Shea Photography; A-66 ©2002 Timothy O'Shea Photography; A-69 Lab-Aids; A-72 David J. Sams/Stock Boston; A-75 Joseph Nettis/Photo Researchers Inc. Part 2 opener: A-80 tl: David Parker/Science Photo Library/Photo Researchers Inc., tm: ©PhotoDisc, tr: Stephen Frisch/Stock Boston, m: Willie Hill, Jr./Stock Boston, bl: Stephen Agricola/Stock Boston, br: ©PhotoDisc; A-81 Harry Wilks/Stock Boston; A-82 Maximilian Stock Ltd./Science Photo Library/Photo Researchers Inc.; A-84 Rosenfield Images/Science Photo Library/Photo Researchers Inc.; A-85 t: Rose Craig; A-89 Bob Daemmrich/Stock Boston; A-91 Lab-Aids; A-94 Frans Lanting/Photo Researchers Inc.; A-102 Stephen Frisch/Stock Boston; A-104 Aaron Haupt/Stock Boston; A-105 Susan Leavines/Photo Researchers Inc.; A-112 Harry Wilks/Stock Boston; A-117 t: ©PhotoDisc, b: A. Ramey/Stock Boston; A-119 Lab-Aids

Materials

B-1, Activity photo/icon: Richard List/Corbis; B-2 tl: ChromoSohm/Photo Researchers Inc., mb: Phyllis Picardi/Stock Boston, br: Peter Menzel/Stock Boston, lm: Will & Deni McIntyre/Photo Researchers Inc., bl: Philadelphia Museum of Art/Corbis, tr: Syracuse Newspapers/The Image Works, mr: David Grossman/The Image Works; B-4 Lab-Aids; B-7 Topham/The Image Works; B-9 tl: Stephen Frisch/Stock Boston; tr: Peter Harholdt/Corbis; bl: Steve Chenn/Corbis; br:Thomas A.Heinz/Corbis;B-10 tl: Mark C.Burnett/Stock Boston; ml: Matthew Borkoski/Stock Boston; bl: Gary Meszaros/Photo Researchers Inc.; tr: Peter Harholdt/Corbis;br: Ted Spiegel/The Image Works; B-12 Jeff Greenberg/The Image Works; B-14 Bob Daemmrich/Stock Boston; B-19 l: B & C Alexander/Photo Researchers Inc.; r: Michael Boys/Corbis;B-20 tl: Ludovic Maisant/Corbis; bl: Mark Godfrey/The Image Works; tr: Roy/Explorer/Photo Researchers Inc.; br: Margot Granitsas/The Image Works; B-22 Lab-Aids; B-25 t: Bettmann/Corbis; b: AFP/Corbis; B-29 Bettmann/Corbis; B-30 Charles Gupton/Stock Boston; B-31 Science Photo Library/Photo Researchers Inc.; B-32 Jeff Greenberg/Photo Researchers Inc.; B-40 Martin Rogers/Stock Boston; B-42 Joseph Sohm, ChromoSohm/ Corbis; B-43 Bettmann/Corbis; B-44 Bill Lane/Pioneer Press; B-45 Wolfgang Kaehler/Corbis; B-51 t: Chad Weckler/Corbis; b: John Wilkes Studio/Corbis;

B-52 t: Richard Hutchings/ Corbis; bl: FK Photo/Corbis; br: Cathy Crawford/Corbis; B-55 Lab-Aids; B-58 Lab-Aids; B-61 Lab-Aids; B-63 t: Hulton Deutsch Collection/Corbis; b: Hulton Deutsch Collection/Corbis; B-65 Lab-Aids; B-67 Aronson/Stock Boston; B-68 Roger Ressmeyer/Corbis; B-69 Jim Sugar Photography/Corbis; B-70 Seventeenth Street Studios; B-73 Tannen Maury/The Image Works; B-76 John DeWaele/Stock Boston; B-81 David Muench/Corbis; B-85 Vince Streano/Corbis; B-86 l: Corbis r: Photodisc; B-92 Kathy McLaughlin/The Image Works; B-97 Eastcott-Momatiuk/The Image Works; B-99 Jeremy Horner/Corbis; B-100 Bob Mahoney/The Image Works; B-105 Science Source/Photo Researchers Inc.; B-111 Lowell Georgia/Photo Researchers Inc.; B-113 Topham/The Image Works; B-114 Bettmann/Corbis; B-124 Bob Rowan,Progressive Image/Corbis; B-125 John Elk III/Stock Boston; B-126 Phyllis Picardi/Stock Boston; B-127 Charles Pefley/Stock Boston; B-128 Bill Ross/Corbis; B-132 l: Jules Frazier/PhotoDisc, ml: Brian Leng/Corbis, mr: Jeremy Hoare/PhotoDisc, r: PhotoDisc; B-133 l: Ken Samuelsen/PhotoDisc m: PhotoDisc, r: Ryan McVay/PhotoDisc; B-134-135 all PhotoDisc.

Energy

Activity photo/icon ©Corbis; C-1 Barth Falkenberg/Stock Boston C-2 all ©Corbis; C-4 ©2002 Timothy O'Shea Photography; C-7 Michael Neveux/Corbis; C-8 ©2002 Timothy O'Shea Photography; C-9 ©2002 Timothy O'Shea Photography; C-11 Underwood and Underwood/Corbis; C-16 Aaron Haupt/Stock Boston; C-18 Bettmann/Corbis; C-19 Martyn Austin/Corbis; C-26 ©2002 Timothy O'Shea Photography; C-28 ©2002 Timothy O'Shea Photography; C-30 Bill Gilletto/Stock Boston; C-31 Raymond Forbes/Stock Boston; C-37 Susan Spann; C-40 ©2002 Timothy O'Shea Photography; C-44 Richard T. Nowitz/Corbis; C-49 Lambert/Archive Photos/Getty Images; C-53 Leonard Harris/Stock Boston; C-60 top to bottom: ©Corbis, ©Corbis, ©Corbis, Warren Gretz/Photo Researchers, Inc.; C-61 t: ©Corbis, m: Steve Harrell/Photo Researchers, Inc., b: ©Corbis; C-67 John Lei/Stock Boston; C-71 Francois Gohier/Photo Researchers Inc.; C-76 Brad Hitz/Getty Images; C-77 Susan Spann; C-80 ©Corbis; C-85 ©PhotoDisc; C-86 PhotoDisc; C-87 Robert Landau/Corbis; C-88 Susan Spann; C-89 Corbis; C-97 Susan Spann; C-101 Ralph Clevenger/Corbis; C-102 John Hulme/Eye Ubiquitous/Corbis

Environment

D-1, Activity photo/icon: David Ulmer/Stock Boston; D-2 tl: Peter Menzel/Stock Boston, ml: Frank Siteman/Stock Boston; bl: Bill Horsman/Stock Boston; tr: Bob Rowan, Progressive Image/Corbis; mr: Danny Lehman/Corbis, br; John Coletti/Stock Boston; D-4 Craig Lovell/Corbis; D-5 t: Dion Ogust/The Image Works, mt: Bob Daemmrich/The Image Works, mb: David Frazier/Stock Boston, b: Stephen Frisch/Stock Boston; D-8 Ed Young/Science Photo Library/Photo Researchers Inc.; D-12 Peter Southwick/Stock Boston; D-13 Eastcott-Momatiuk/The Image Works; D-14 Kevin R. Morris/Corbis; D-15 Ron Dorsey/Stock Boston; D-16 t: Chris Knapton/Science Photo Library/Photo Researchers Inc., b: Aaron Haupt/Stock Boston ; D-17 Vince Streano/Corbis; D-18 Steve Weinrabe/Stock Boston; D-19 Jenny Hager/The Image Works; D-21 David Ulmer/Stock Boston; D-22 Dean Abramson/Stock Boston; D-28 l: Bob Daemmrich/Stock Boston, r: Vittoriano Rastelli/Corbis; D-29 John Coletti/Stock Boston; D-30 Chinch Gryniewicz/Ecoscene/Corbis;D-33 David Ulmer/Stock Boston; D-34 t: Tony Savino/The Image Works, b: Galen Rowell/Corbis; D-35 Bob Krist/Corbis; D-36 tr: C.J.Allen/Stock Boston, mr: Syracuse Newspapers/Carl Single/The Image Works, l: Danny Lehman/Corbis, br: Garry D.McMichael/Photo Researchers Inc.; D-46 George Dineen/Photo Researchers Inc.